Happy Birthday Dad: enjoy

Wootten

of Wenhaston

[illegible] + Chris x.

The Plantsman's Handbook

Blackheath, Wenhaston, Halesworth, Suffolk. IP19 9HD
Telephone: 01502 478258
Fax: 01502 478888
e-mail: info@woottensplants.co.uk
www.woottensplants.co.uk

Published in 2013 by Woottens Plants
ISBN 978-0-9927764-0-4

Design by www.wearedrab.net

Acanthus mollis *(3)*

Introduction to the Plantsman's Handbook

My brother Michael was a man of strong opinions, which he firmly believed were entirely rational – a family characteristic. As one of five articulate siblings he was used to vigorous argument, and life had given him much material. Michael had studied Russian literature and language at Essex and London Universities and spent a miserable winter in Minsk (streets littered with vodka-swilling drunks) before settling in rural Norfolk, to make goats' cheese. Then it was off to London, where he worked at the fledgling business of Neal's Yard Wholefood Warehouse - which he eventually bought and turned into an extraordinary enterprise. Innovative methods of employment and an exceptional range of products were complemented by Michael's somewhat idiosyncratic attitude to marketing - he commissioned Tim Hunkin to make a water-clock for the outside of the building, which drenched unwary passers-by whenever it struck the hour.

But eventually his love of plants and gardening, learnt from our mother, brought him back to Suffolk, where he established Woottens of Wenhaston in 1991. It soon became famous, partly for the extraordinary health and quality of the plants he raised in his nursery, partly for the way that he described and presented them. As Anna Pavord commented: "The maverick in him is much more interesting. In deciding what to grow he mostly follows his own nose, each year propagating hundreds of new plants just because he wants to see what they are like." Such experiments led to passionate enthusiasms and equally vehement dislikes – prompted by Michael's strong aesthetic sense but also because plants, for him, were endowed with moral attributes; some were devils, others angels.

It was those opinions, so vigorously expressed, which gave irresistible character to his wonderfully written catalogues. Treasures of beautiful production, fine illustration and vivid description, these handbooks soon became famous - collected by customers, rivals and other writers on gardens and horticulture. Each edition reflected his latest enthusiasms, enriched with idiosyncratic snippets of information, culled from all manner of sources. Eccentric, passionate, erudite – they echo with his voice.

Michael was never satisfied, but the handbook that came closest to his platonic ideal of perfection was this one, from 2005. You will find your own favourite passages as you browse each page and paragraph. My own, the one that most vividly reminds me of Michael, reads as follows.

'Zantedeschias are named after the 19th Century botanist, Giovanni Zantedeschii. They are native to South Africa. My father used to call them pig lilies because in Ireland, arum lilies and pigs are to be found nuzzling up to each other in muddy ditches. He did not like them. I think he thought, by claiming the name of lily, they blasphemed against true lilies, such as Lilium candidum, which for him was a sacred thing, in all its trumpet glory, broadcasting the Annunciation eternally. For me Zantedeschias have their voluptuous appeal, the absolute white nakedness of their spathes against the lush deep leaves.'

Of course I disagree with Mike (sharing my father's dislke of Arum lilies) but that's an argument we can no longer enjoy together. I turn instead to his catalogue and lose myself with pleasure in byways of botanical erudition, relishing Michael's enjoyment of language, his delight in ephemeral things, flowers that bloom and fade. But I also have a sense of permanence, of continuity, for the Plantsman's Handbook celebrates and enriches a great English tradition of horticultural writing, centuries old. You have in your hands a classic.

Simon Loftus – April 2013

Contents

Herbaceous Plants for Sun 3

Hardy Geraniums for Sun 140

Hardy Geraniums for Shade 148

Grasses for Sun 150

Grasses for Shade 162

Herbaceous Plants for Shade 164

Ferns 200

Hostas 202

Credits 206

Plants for Problem Areas 206

Bibliography 208

Agapanthus africanus Mrs P Loftus *(2)*

Herbaceous Plants for Sun

ACAENA. (Rosaceae). Acaenas are evergreen mat forming prostrate plants with pinnate foliage and burr like seedheads. Due to their vigorous habit they should be planted with caution and be kept away from more dainty and delicate subjects. Can look good in troughs or pots. Any well drained soil. **A. *buchananii*** has bluish grey foliage. Nondescript flowers. June-July. Ht. 10cm. Sp. 25cm.

ACANTHUS. (Acanthaceae). Acanthus The English names Bears Breeches or, as the Tudor writer Turner quaintly called it, Branke Ursine (names for which no satisfactory explanation has ever been provided) do not do justice to the dignity of this species. Shaped like a huge snapdragon, Acanthus is one of the great architectural beasts of the garden. **A. *mollis*** puts up tall spikes of purple and white flowers. Jul-Oct. Its luxuriant, large, basal leaves are a rich, dark, glossy green and deeply cut. They are evergreen in a mild winter. Ht. 120cm. Sp 70cm. **A. *m. Hollard's Gold.*** The leaves of this variety are at their best in winter when they are bright golden green. Good planted with Bergenia Eroica. Ht. 90cm. Sp. 70cm. **A. *m.* Alba.** A form with pure white flowers and bracts. **A. *spinosus*** has spiny, deeply divided, mat green leaves and is deciduous. Its flowers are very similar to A. mollis but are borne on a shorter stem. Ht. 120cm. Sp. 70cm. **A. *hungaricus*** is shorter than A. spinosus and has less divided foliage. Ht. 100cm. Sp. 70cm.The leaves of A. spinosus are those depicted on the capitals of Corinthian columns. Acanthus spinosus is a dependable flowerer; Acanthus mollis only flowers after a warm winter. If Acanthus mollis has its foliage cut down by frost in winter, it will refuse to flower the following summer. Both plants need lots of space and look good spilling out onto gravel. Despite the height of their flowering stems, Acanthus should not be planted at the back of the border, as without visible foliage they are but sorry things. Spent flowering stems are decorative, decked with large green seed-pods, which gradually fade to brown, splitting to reveal single, awesome, shiny, black seeds. Acanthus need careful siting, as once established, their roots are impossible to eradicate. Always plant Acanthus in full sun and well-drained soil. Planted in shade they become afflicted with mildew and waterlogged in winter they rot and die. All the mentioned Acanthus are drought tolerant once their taproot is established. They make good seaside plants and are resistant to salt. Acanthus do not get eaten by deer; slugs and snails are not particularly keen; rabbits may sometimes show an interest. All forms of Acanthus are readily propagated from root cuttings in early spring. Seed can be germinated under gentle heat in April. Acanthus mollis has an invasive habit and is therefore more suitable for the wilder parts of the garden; it can also look good in large containers. Acanthus hungaricus is a liberal selfseeder, which makes it undesirable for the small garden but useful for landscaping. A. spinosus is the least thug-like and therefore the best adapted for planting in the mixed border. Acanthus look good planted in combination with other bold architectural plants: Astelias, Phormiums, Geranium palmatum, Stipa gigantea, the larger Eryngiums, all manner of Echinops, Erysimum Bowles Mauve, Crambe cordifolia, Crambe maritima, Salvia sclarea var. Turkestanica. E. A. Bowles planted a whole bed at Enfield House with Acanthus and Crambe cordifolia. Acanthus also looks at ease with fellow Mediterraneans, Cistus and Rosemary and is an essential plant for the dry or gravel garden.

ACHILLEA. (Asteraceae). The Achillea family gains its name from the Greek hero Achilles, who is said to have discovered it "and with this same wort healed them who were stricken and wounded with iron", hence its vernacular name, Bloodworte. In Celtic culture, Achillea was valued by women. This Celtic poem recorded by Kevin Jackson in 'Early Celtic Poetry 1935' shows that eating the plant by women was thought to illumine their femininity and sexuality: "I will pick the smooth yarrow that my figure may be sweeter, that my lips may be warmer, that my voice may be gladder. May my voice be like a sunbeam, may my lips be like the juice of the strawberry. May I be an island in the sea, may I be a hill in the land, may I be a star in the dark time, may I be a staff to the weak one. I shall wound every man, no man shall hurt me". **A. *millefolium***, our native Yarrow, is sun loving, totally drought tolerant, and will grow in pure sand. It has slightly rounded heads of close packed, small flowers. Cut back midseason to encourage repeat flowering. **A. *m. Summer Wine*** has deep, burgundy red flowers. June-Aug. Ht. 60cm. Sp. 40cm. **A. *m. Red Velvet*** is, I think, the best red cultivar yet. The flowers have great intensity and retain their colour longer than most. June-Sept. Ht. 60cm. Sp. 40cm. **A. *m. Rose Madder*** is a useful short cultivar with, according to Bob Brown, the longest flowering period. Soft pinky red flowers June-Nov. **A. *ptarmica The Pearl*** will grow in sun or part shade; a moisture loving plant, it thrives in clay or on the margin of ponds and streams. Loose sprays of pure white, small, double flowers. June-Aug. Good for flower arranging. Ht. 70cm. Sp. 50cm. Spreading. If chewed, this plant is supposed to ease toothache. Its young shoots are good in salads. In the Highlands of Scotland, the wild, single form of this plant was dried and used as a substitute for snuff; hence one of its common names Sneezewort Yarrow. Like many plants with small pompon flowers, this Achillea has colloquially been called Bachelors Buttons. A. p.The Pearl can be a bit weak on its legs and **A. *p. Nana Compacta***, a dwarf compact form with single flowers is to be preferred for edging ponds and streams. Plant with the yellow flowered Caltha palustris flore pleno – the Achillea will start flowering after the Caltha has ceased and the Caltha's bold foliage will makes a handsome contrast to the Achillea's filigree. Ht. 30cm. Sp. 40cm. **A. *p. Stephanie Cohen*** has delicate mauve pink flowers and should be planted with Caltha palustris Alba, which unlike the yellow form repeat flowers into the summer. The following Achilleas all derive from the Caucasian plant **A. *filipendulina*** and have large flat heads, which make a good contrast with feathery grasses; they require rather better, more moisture-retentive soil than the A. millefolium hybrids. Christopher Lloyd is predictably rude about these hybrids, mocking them as pandering "to the tastes of the pastel shades brigade, haters of flowers that are bright yellow, orange or scarlet." **A. *Anthea*** has flat heads of light yellow flowers. June-Aug. Ht. 60cm. Sp. 60cm. **A. *Coronation Gold*** has rich yellow flowers and grey green leaves. June-Sept. Ht. 60cm. Sp. 40cm. **A. *Fanal*** has flat heads of terracotta orange flowers June-Aug. Ht. 60cm. Sp. 60cm. **A. *Feuerland*** bears bright red flowers which fade to apricot. July-Sept. Ht. 100cm. Sp. 60cm. **A. *Forncett Fletton*** has brick orange flowers. June-Aug. Ht. 75cm. Sp. 60cm. **A. *Inca Gold*** has flat heads of terracotta orange flowers. June-Aug. Ht. 60cm. Sp. 60cm. **A. *Marie Ann*** is a useful dwarf form with pale, lemon yellow flowers. June-Sept. Ht. 40cm. Sp. 30cm. **A. *Martina*** has palest yellow flowers with a touch of green. Very refined and flowers over an extended period. June-Oct. Ht. 50cm. Sp. 40cm. **A. *Terracotta*** has warm terracotta flowers. June-Aug. Ht. 80cm. Sp. 40cm. **A. *Walther Funcke***, selected by the great plantsman, E Pagels, is a wonderful fiery red. June-Aug. Ht. 60cm. Sp. 30cm.

Agapanthus umbellatus *(3)*

AEONIUM. (Crassulaceae). Large succulents native to the Canaries, Aeoniums can grow into small trees. Not hardy, but easy in a large light window. Good bedded out in summer with Agaves, Astelias, Cannas, Echeverias, Musa Basjoo and suchlike exotica. Drought tolerant. They will all grow to a height of 150cm. with a spread of about 40cm. A. ***arboreum Schwarzkopf*** has dramatic, dark bronze, almost black foliage. This variety can become leggy. To keep compact, carefully grind out, with your small fingernail, the middle of the top rosette each spring. This will encourage good side growth. A. ***arboreum Magnificum*** is self-branching and has plain green leaves; it is perhaps the most architectural variety.

AGAPANTHUS. (Amaryllidaceae). Agapanthus, despite their misleading English common name, Lily of The Nile, are native to South Africa not Egypt. They have long been cultivated in this country; the first Agapanthus to have flowered in England was recorded at Hampton Court in 1697. All the Agapanthus listed below (unless otherwise stated) are easy garden plants, hardy to at least - 15°C. and needing no special protection. They like full sun and rich moist soil. The white and blue form of **A. *campanulatus***: **A. *c. Albidus*** has white flowers borne on brown pedicels July-Oct. Ht. 60cm. Sp. 40cm. and **A. *c. Navy Blue*** has dark blue flowers Aug-Sept. Ht. 60cm. Sp. 40cm. **Hybrid Agapanthus. A. *Back In Black,*** a choice variety, has jet black stems which bear dark purple blue flowers which open from near black buds. July-Aug. Ht. 65cm. **A. *Black Panther***, another variety, bears full heads of blue black flowers on strong stems July-Sept. Ht. 65cm. Sp. 40cm. **A. *Blue Heaven.*** Tender. Mid-blue flowers June-Sept. This is the only truly remontant Agapanthus I know. Ht 90cm. Sp. 40cm. **A. *Bressingham White*** is for me the best white. Flowerbuds have a hint of blue but the open flowers are pure white. July-Sept. Ht. 90c. Sp. 50cm. **A. *Gayle's Lilac*** is a short growing form with lilac flowers Aug-Oct. Ht 40cm. Sp. 40cm. **A. *Jack's Blue***, my favourite Agapanthus, a wonderful introduction from New Zealand has huge dense heads of rich deep purple blue flowers over an extended season. July-Oct. Ht. 150cm. Sp. 40cm. **A. *Pinocchio.*** Cobalt blue flowers. July-Aug. Ht. 60cm. Sp. 30cm. **A. *Loch Hope*** has very large heads of mid blue flowers July-Sept. Ht. 1.2m. Sp. 40cm. **A. *Midnight Blue*** has deep navy blue flowers July-Sept. Ht. 90cm. Sp. 50cm. **A. *New Blue*** Deep blue flowers. Individual flowers exceptionally Large. Late flowering. Aug-Oct. Ht. 75cm. 40cm. **A. *praecox Flore Pleno.*** Large heads of double, pale powder blue flowers. July-Sept. Ht. 50cm. Sp. 50cm. Broad leaves. Tender. Wonderful in tubs. **A. *inapertus*** has tall sturdy stems and is useful for its late flowering. Its flowerheads are not spherical rather gracefully drooping somewhat in the manner of Allium pulchellum var. carinatum or of Allium flavum. A. inapertus is reliably hardy. **A. *i. Sky*** has light blue flowers Aug-Sept. Ht. 80cm. Sp. 40cm. **A. *i. Purple Cloud*** has superb, deep purple flowers Aug-Sept. Ht. 80cm. Sp. 40cm. **A. *Snowdrop*** has dense heads of white flowers July-Sept. It has distinctive yellow flowering stems. Ht. 50cm. Sp. 40cm. **A. *Timaru*** has dense heads of vivid blue flowers over a long period. July-Oct. Ht. 75cm. Sp. 40cm. **A. *africanus*** is a very variable plant. Two hardy forms **A. *a. Blue***, and **A. *a. Albus*** are both about 90cm. tall with relatively modest heads of flowers. A much bigger tender form, **A. *a. Mrs. P. Loftus*** is spectacular at all seasons of the year, with broad, evergreen, strap-like leaves, huge umbels of rich blue flowers July-Sept and magnificent architectural seed heads through into the winter. This Agapanthus is for me perhaps my absolute favourite plant for large pots and tubs. Ht. 120cm. Sp. 70cm. All Agapanthus thrive on congestion and container grown plants should only be divided infrequently.

Allium christophii *(2)*

They resent, however, being pot bound, (pot bound plants will be reluctant to flower) and should be repotted annually in March. During the growing season they require plentiful moisture and regular feeding with tomato fertiliser or some other high potash feed. For those with a passion for all things variegated there is a dwarf form called **A. *Tinkerbelle.*** Never flowers for me, although I have seen a photo of it flowering in the Canaries! Grown for its evergreen, narrow, silver variegated foliage. Very pretty in a pot. Not hardy, but decorative in the conservatory in winter. Ht. 15cm. Sp. 25cm.
AGASTACHE. (Lamiaceae). All Agastaches have aromatic foliage and are resistant to browsing by deer. Their flowers are much frequented by bees and butterflies. **A. *astromontana Pink Pop*.** Dwarf compact plant. Spikes of tubular, scented pink flowers. May-Nov. Ht. 28cm. Sp. 30cm. **A. *foeniculum Golden Jubilee*.** Bright gold foliage with spikes of tubular, scented blue flowers. Compact. Ht. 50cm. Sp. 40cm. **A. *mexicana*** has salvia like flowers and grey green foliage. All varieties flower June-October and need sharply drained soil and full sun. They are dainty, refined plants and have a somewhat lax habit, which looks at home in the gravel garden. They all have a spread of about 40cm. **A. *m. Apricot Sprite*** has heads of large deep reddish apricot flowers. Ht. 50cm. **A. *m. Painted Lady*,** a sport from A. Firebird, has soft, pink flowers. Ht. 75cm. **A. *m. Tangerine Dreams*** has vivid orange flowers. Ht. 60cm. **A. *rugosum Alabaster*** is a much more robust plant with upright stems and heads of densely packed small white flowers from June-September. **A. *urticifolium Blue Wonder*** has a similar habit to A. r. Alabaster and has upright stems of purplish blue flowers in July-Sept. Ht. 70cm. Sp. 40cm. Both these varieties are native to North America. Both look good with grasses and are strong structural plants. They require sun and moistish soil.
AGAVE. (Asparagaceae). **A. *americana Variegata*.** Utterly vicious and wonderful. The thickest of leather gardening gloves are necessary for handling this wonderful beast. Not hardy. But looks as splendid in the conservatory in the winter as it does on the terrace in the summer. Ht. 60cm. Sp. 60cm.
ALCEA. (Malvaceae). Hollyhocks are the quintessential, cottage garden plant but, sadly, they are more often seen on chocolate boxes than in gardens. This is due partly to their great bulk, which is difficult to accommodate in a small garden and partly to availability. Nurserymen rarely grow them as they are troublesome in pots. Hollyhocks are not difficult to grow in the garden. They need full sun and prefer a not too dry, moisture retentive soil. A constant supply of moisture will enable them to better withstand the ubiquitous Hollyhock Rust. In old gardens they were often positioned so their roots would run under paving, which keeps them cool in hot weather and reduces stress. Stress induced by drought or overheated roots greatly exacerbates problems with rust. Hollyhocks, which are not stressed, will grow through rust and most professional gardeners have now stopped spraying them with fungicides, as this is an endless treadmill. To ensure robust plants with plentiful flowers, it is a good idea to stop the main stem, thereby encouraging the lateral growths to develop more fully, and if these in turn are stopped, more bushy, less gangly plants will be achieved. One of the best uses of hollyhocks, I have seen, is in an old cottage garden, where they are planted in informal lines on either side of a broad path. Goethe, I have read, planted a veritable avenue of hollyhocks at his house at Weimar. When in flower, friends would be invited to drink tea and to admire. Ah, the vanity of gardeners! John Lawrence writing in 1726 considered the siteing of Hollyhock to be critical. He wrote; "proper places against walls or the

Allium karataviense *(1)*

Corners of gardens should be assigned to them, where they may explain their beauty to distant views." The doubles seem to me mere baubles, rosettes for horsey girls to win at pony club. **A. *rosea Nigra*** is a much lusted after, nearly black form. July-Sept. Ht. 175cm. Sp. 50cm. **A. *rugosa*** is a delicious, pale yellow flowered variety from the Caucasus. It is more resistant to rust than A. rosea. Ht. 120cm. Sp. 40cm. **A. *ficifolia***, rather more rust resistant and slightly shorter than A. rosea, all plants will have single flowers. Pot-luck as to colour. July-Aug. Ht. 150cm. Sp. 70cm.

ALLIUM. (Amaryllidaceae). All members of the allium family thrive in any well drained soil. Most are drought tolerant. They look good in the gravel garden. All alliums look better planted in informal drifts.

Taller varieties. A. *azureum* bears umbels of sky blue flowers in June. Ht. 50cm. Sp. 20cm. **A. *christophii*** has huge, silvery blue inflorescences (20cm. in diameter) May-July. Noble seed heads persist well into the autumn, if not prematurely immolated for a dried flower arrangement. Thrives in all soils including London clay. Ht. 40cm. Sp. 25cm. **A. *nigrum*** has flat heads of white flowers with a green tinge and black centres. Strong growing. June-July. Ht. 40cm. Sp. 30cm. **A. *siculum* syn. Nectaroscordum siculum.** Soft green and purple flowers. One of those flowers Christopher Lloyd delights in sneering at for being insufficiently garish. Good for flower arranging. June. Ht. 90cm. Sp. 20cm. **A. *sphaerocephalum*** has reddish purple drumstick flowers and is valuable for its late flowering. Looks good with Artemisia Lambrook Silver. July-Aug. Ht. 60cm. Sp. 20cm. **A. *spherocephum Hair*** is the floral equivalent of a bad hair day. A disarray of thin green filaments takes the place of a flower. Flower arrangers love it. Sterile. July-Aug. Ht. 60cm. Sp. 15cm.

Shorter Varieties. A. *carinatum var. pulchellum*. Heads of soft lilac pendant bells. Like A. flavum, very informal and easy to use. Flowers Aug-Sept. Looks good growing through Artemisia schmidtiana. Increases rapidly. Ht. 35cm. Sp. 30cm. **A. *geyerii*.** Umbels of pink bell shaped flowers. Late summer through into autumn. Ht. 30cm. Sp. 30cm. **A. *c. pulchellum Album*** is the lovely and inexplicably rare white form. **A. *neapolitanum*** has white flowers in May-June. Ht. 50cm. Sp 25cm. **A. *paradoxum var. normale*** has gleaming white flowers. May. Fresh green leaves. Ht 30cm. Sp. 25cm. Good in heavy soil. Self seeds. **A. *senescens var. glaucum*** has pink flowers July-Sept. Decorative twisted bluish leaves. Thrives in dry rocky places. Central Asia and Europe. Ht. 25cm. Sp. 20cm. **A. *schoenoprasum Forescate*** is a selected form of chives with larger, cheerful, reddish rose flowers. June-July. Ht. 25cm. Sp. 25cm. Lovely as an edging plant. The flowers dry well - if you must. **A. *flavum*** has sweetly scented, pure gold, pendant flowers. July-Aug. Ht. 30cm. Sp. 15cm. **A. *oreophilum*** has miniature, shocking pink flowers in May-June, followed by attractive, white seed heads. Ht. 15cm. Sp. 25cm. **A. *moly* syn. luteum** has bright yellow flowers. June. Ht. 30cm. Sp. 40cm. Loves lime and multiplies rapidly. Both A. moly and A. flavum provide good summer colour on the rockery, if you have not dynamited it in the cause of good taste.

A. *karataviense* has spectacular, glaucous winter foliage and large white flowers tinged with pink, (20cm. in diameter). Christopher Lloyd correctly describes them as not very spectacular but all must be forgiven for their delicious fragrance. May-June. Ht. 20cm. Sp. 25cm.

ALOE. (Asphodelaceae). **A. *Black Gem*.** Dark bronze spiky succulent. Orange red flowers in spring. Ht. 12cm. Sp. 15cm. Looks good with Sempervivums. Protect from heavy frost.

Alstroemeria aurantiaca Lutea *(2)*

ALSTROEMERIAS. (Alstroemeriaceae). The first Alstroemerias were introduced into England from South America in 1754. Alstroemerias are easy, vigorous plants. They thrive in any soil except heavy clay. Happy in sun or part shade. All the following Alstroemerias are hardy. They resent root disturbance and should, when lifted for splitting, be replanted in large clumps. **A. *pulchella* syn. A. psittacina,** the parrot lily, whose striking red and green flowers are much sought after by flower arrangers, July-Nov. Ht. 60cm. Sp. 60cm, **A. *aurantiaca Lutea*** has yellow flowers. **A. *aurantiaca*** has fiery orange flowers. Both flower July-Aug. Ht. 80cm. Sp. 60cm. The following varieties of Parigo Alstroemerias, bred by John Goemans, are remarkable plants flowering from June to November with good strong upright stems. Lovely for picking. **A. *Flaming Star*** has brilliant orange flowers and a strong upright habit. **A. *Orange Glory*.** Orange flowers. Vigorous habit. Ht. 90cm. **A. *Purple Rain*** has rich purple flowers. **A. *Red Beauty*** has red flowers and a compact habit. **A. *Selina*** has pink flowers. **A. *Spitfire*** has scarlet red flowers, its foliage neatly edged with lime green. All grow to about 90cm. tall. **A. *White Apollo*** has pure white flowers with yellow blotches in the throat. While the Parigo hybrids are definitely plants for the border (too tall to be useful in pots) the Princess hybrids seldom exceeding 30cm. in height are ideal for container growing. Like the Parigo hybrids the Princess hybrids give a constant display throughout the summer and autumn. **A. *Princess Daniella*** has cream flowers. **A. *Princess Ivana*** has purple pink flowers. **A. *Princess Stephanie*** is a pink and white bicolour. **A. *Princess Zavina*** has salmon orange flowers.
ALTHAEA. (Malvaceae). **A. *officinalis*.** Althaea officinalis (Marsh Mallow) is native to the salt marshes of the Thames estuary, "an exquisite surprise in its muddy, salty, desolate, smelly surroundings". (Geoffrey Grigson). Spires of small pale pink flowers rise on narrow stems to 180cm. but miraculously need no staking May-Sept. **A. *o. Romney Marsh*.** This selected form is much more compact than the species with a height of only 40cm. Sp. 35cm. It has velvety leaves and white, pink flushed, smallish flowers May-Sept. Both these plants are rust free. The species plant looks superb with Lythrums, Eupatoriums and Valerian officinalis. They will grow in any moist or boggy soil. Before marshmallows fell into the hands of the confectionery corporations, they were made from the roots of Althaea officinalis. Nowadays they are a concoction of gelatine, sugar and starch.
ALYSSUM. (Brassiaceae). **A. *saxatile*.** Central European grey-leafed subshrub, smothered with golden yellow flowers from May to July. Ht. 25cm. Sp. 90cm. Ideal for clothing a dry sunny bank.
AMSONIA. (Apocynaceae). Amsonias gained their name from Dr Charles Amson an 18th Century scientist, who travelled in North America. **A. *tabernaemontana*** is widespread throughout the South East states of the U.S.A. It bears umbels of pale blue flowers, (I think they are exquisite, Christopher Lloyd calls them wishy-washy) in May-July, borne on sturdy but delicate stems. Its narrow lanceolate leaves turn a brilliant yellow in autumn. Thrives in any not too dry soil. Sun. Lovely planted with Anthericum ramosum.
ANCHUSA. (Boraginaceae). ***Anchusa azurea* syn. italica** was introduced from Italy in 1810. **A. *a. Loddon Royalist*** has the purest, deepest blue flowers of any plant I know - better even than Salvia patens - May-July and **A. *a. Opal*,** which has pure, Cambridge blue flowers – June-August. Both grow about 90cm. tall and need rich soil and sun. Spread approximately 30cm. Anchusas should be staked before flowering as they have a

Alstroemeria Purple Rain *(2)*

tendency to recumb. Neither variety is long lived, but they can easily be renewed from root cuttings. Christopher Lloyd likes Anchusa flowers sprinkled on his salad.
ANEMONE. (Ranunculaceae). **A. x *lesseri*.** Hybrid between **A. *multifida*** and **A. *sylvestris*.** Attractive deeply cut foliage. Carmine red flowers. Good in gravel where it will self seed. May-June. Ht. 25cm. Sp. 20cm. Sun. Well drained soil.
ANGELICA. (Apiaceae). All Angelicas are biennial but reliably self seed in moist soil. They all dislike poor dry sand. **A. *acutiloba*.** Finely cut, glossy, deep green foliage and heads of white flowers Jun-Jul. Ht. 60cm. Sp. 40cm. I am told this plant will behave as a perennial if the flower heads are removed before seed is set. Moist soil. **A. *Corrine Tremaine*** has theatrical large cream splashed leaves. Cream and green flowerheads. June-Aug. Ht. 60cm. Sp. 50cm. **A. *gigas*.** A magnificent if rather fashionable plant. Large divided, deep green leaves. Purple stems. Dense, domed beetroot flowers. June-Aug. The flowers strongly attract flies. - dark red, covered in flies, they resemble raw chunks of well hung beef which somebody has forgotten to put back in the fridge! Ht. 140cm. Sp. 50cm. Angelica gigas looks good planted with other black plants: Alcea rugosa Nigra, Actea simplex Brunette, Scabiosa atropurpurea. **A. *sylvestris Vicar's Mead*.** Black leaves and stems. Pink flowers. Jun-Sep, 75cm. Sp. 30cm. Stunning form of a British native.
ANISODONTEA. (Malvaceae). **A. *capensis*.** Produces a profusion of deep pink flowers throughout the year in a sheltered position. Glossy dark green leaves. Evergreen. Ht. 100cm. Sp. 70cm. Sun or part shade. Not for very frosty gardens. Well drained soil. ***A. c. Tara's Pink*.** Profuse shell pink flowers throughout the year. Best against a South or West wall. Grey green leaves. Ht. 250cm. Sp. 70cm. Sun or part shade.
ANTHEMIS. (Asteraceae). **A. *carpatica Karpatenschnee*** bears white flowers from May to July and makes dense ground cover. Much the tidiest and neatest Anthemis Ht. 15cm. Sp. 30cm. Needs sharp drainage and sun. **A. *cupaniana*** has intensely silver, filigreed foliage, which is covered in chalk white daisy flowers throughout May and June. Ht. 38cm. Sp. 60cm. Good on a hot sunny bank, making as it does "a vast decumbent mass". (Reginald Farrer). Christopher Lloyd suggests using A. cupaniana to underplant tall late flowering tulips. **A. *tinctoria*.** First described by John Goodyer in 1621, who referred to the flower of the species plant as a "great yellowe ball or dish". All cultivars have daisy flowers. Their spring foliage looks edibly fresh. All are easy, sun-loving plants, which will flower throughout the summer, given sufficient moisture. They need dividing every couple of years or they will repine. Replant into fresh soil. **A. *t. E. C. Buxton*,** has cool, lemon yellow flowers, ferny green foliage and is exceptionally long flowering. June-Nov. Ht. 50cm. Sp. 50cm. G. Stuart Thomas says "that a summer border without A. t. E. C. Buxton is difficult to conceive". He recommends planting it with Campanula lactiflora. **A. *t. Sauce Hollandaise*** has very pale, creamy white flowers and ferny green foliage. June-Sept. Ht. 38cm. Sp. 50cm. **A. *t. Susan Mitchell*** has large, creamy yellow flowers with silvery foliage. June-Sept. Ht. 50cm. Sp. 50cm. **A. *t. Sanctis Johannis*.** Calendula-orange, daisy flowers. June-September. Deep green, filigreed foliage. Ht. 50cm. Sp. 50cm. ***A. t. Tetworth*** has silver foliage and pure white flowers all summer. June-Sept. Ht. 50cm. Sp. 50cm. **A. *t. Wargrave*** has pale, creamy yellow flowers and ferny green foliage. June-Sept. Ht. 50cm. Sp. 50cm.
ANTHERICUM. (Asparagaceae). The name Anthericum is singularly undescriptive. The R.H.S. Dictionary would have us believe that it derives from 'anthericos', Greek for

Alyssum saxatile *(3)*

the stem of the Asphodelus! A. T. Johnson, on the other hand, cites as its origin 'anthos', Greek for a flower and 'therikos', Greek for hedge. Anything less hedge like is hard to imagine! Parkinson referred to Anthericums as Spiderworts for their supposed efficacy in curing the poison of Scorpions. Gerard also used the name Spiderwort for which he found credibility in a supposed visual similarity between anthericums and spiders. Both **A. *liliago*** and **A. *ramosum*** are the airiest creatures imaginable, the elfin sprites, one might say, of the lily family. A. liliago, the St Bernard Lily is perhaps the more imposing of the two. It makes handsome rosettes of silvery green foliage, from which in May-June rise 90cm. spikes of white, trumpet-shaped flowers. Sp. 30cm. Farrer commented on the naming of A. liliago after "the odious St Bernard", that such a thing as that, "so pretty and innocent is far too good for the preacher of the Crusades." Quite right! Anthericum ramosum grows to only 60cm. and unlike A. liliago is much branched. It has grassy foliage and its flowers are numerous and starry, a great delight in the gravel garden from early June till late September. Both Anthericums seem equally at home in both alkaline and acid soil, and will grow in light shade as well as sun. The only soil they will not tolerate is heavy clay or dry sand.

ANTIRRHINUM. (Plantaginaceae). English names: Snapdragon, Lion's Mouth, Calves Snout. The last name, I think, is the most evocative. **A. *braunblanquettii.*** A perennial, hardy, species Snapdragon. Beautiful clear, soft yellow flowers from June to October. In time becomes woody, but reliably renews itself by self-seeding. An inexplicably neglected plant of tremendous beauty, far outshining all those F.1. hybrids. Ht. 45cm. Sp. 30cm. John Hill writing in 1757 says of the Antirrhinum. "Though not a native of our Country, bears the free air perfectly well in it." He also says, correctly I think, that in its unaffected beauty, "it exceeds many of the most pompous flowers." Miller in his Gardener's Dictionary writes. "These plants will grow among stones or the joints of old walls, where they may be placed so as to render some abject part of the garden very agreeable."

AQUILEGIAS. (Ranunculaceae). Both the Latin name Aquilegia ('aquila', Latin for eagle) and the English name Columbine ('columba', Latin for dove) derive from the resemblance of the Aquilegia's two long spurs to a pair of bird's wings. Somehow I think, modest pretty things that they are, they associate better with doves than eagles. Aquilegias are a great resource for our dry, inhospitable sand. All flower May-June. All Aquilegias are liable to very damaging attack by aphids, particularly in April and September. Spraying with an aphicide, containing the systemic chemical Pirimicarb, will control aphids without damaging any beneficial predators. The application of the spray must take place at the very first sign of damage. Aphids can skeletonise an Aquilegia in a matter of a few days. Aquilegias are notoriously promiscuous; different species and varieties wantonly hybridise. Never expect seedlings in your garden to run true. But most progeny will be pretty and worth the raising.

Aquilegias for sun. These Aquilegias are alpines in origin and need a sunny open position and good drainage. Aquilegias do not create a large basal rosette and can be slipped in between later flowering perennials – a gap 20cm. in diameter is sufficient **A. *atrata*** is native to the European Alps. Very dark purple, almost black flowers. May-June. Ht. 50cm. **A. *chrysantha Yellow Queen*** has single, golden yellow flowers with spectacular spurs. May-July. Very decorative, much divided foliage. Ht. 90cm. **A. *flabellata Nana Alba*** has single, ivory white flowers of great substance. May-July. Handsome rosettes of glaucous foliage. Ht. 20cm. **A. *skinneri*** has single red and yellow

Anchusa italica *(3)*

flowers. May-Sept. A unique plant, for the length of its flowering. Decorative, highly divided foliage. Ht. 30cm. Lastly three hybrids: **A. *Crimson Star*** which has carmine pink and white flowers Ht. 60cm. **A. *Silver Queen*,** which has huge white long spurred flowers Ht. 60cm. and **A. *Mrs M Nicholls*** which has blue and white flowers. Ht. 80cm. All three flower late June-July after the A. vulgaris hybrids have finished.

Aquilegias for shade. Theses are listed in the shade section of the handbook.

ARENARIA. (Caryophyllaceae). **A. *montana.*** A fast growing plant which makes huge sheets of deep green foliage. In June covers itself with large, pristine, white flowers. One of the very best plants for a dry sandy bank. Prostrate. Ht. 10cm. Sp. 30cm.

ARGEMONE (Papaveraceae). Showy bright yellow poppy flowers and handsome glaucous prickly foliage. Ht. 100cm. Sp. 40cm. June-July. Thrives in poor dry soil where it will self seed and naturalise itself.

ARGYRANTHEMUM (Asteraceae). The older varieties of Margarites verged on the thuggish. These newer varieties have had bred into them a much more dainty habit. With dead heading they will flower from April to October. All varieties of Argyranthemum require overwintering in a greenhouse. **A. *frutescens Blanche Petite*.** White flowers. Very compact. Ht. 30cm. Sp. 25cm. **A. *f. Julieanne*.** Vibrant red flowers with double centres which fade slowly to pink. April-Oct. Ht. 50m. Sp. 40cm. **A.*f. Sulphur Petite*.** Sulphur yellow flowers. April-Oct. Glaucous foliage. Ht. 40cm. Sp. 30cm. **A. *f. Summer Melody*.** Small soft pink pompom flowers in profusion. Ht. 50cm. Sp. 40cm.

ARMERIA. (Plumbaginaceae). English names: Thrift, Ladies' Cushion, Sea Pink. ***A. maritima.*** Ann Pratt suggests that thrifts gained their English name from their ability to subsist on little, their native habitat being rocks and the cliff edge. The island of Annet in the Scillies is carpeted in its entirety with sea pinks. Armerias are the easiest of plants, thriving in the poorest of dry sandy soils and, of course, tolerant of any amount of salt spray. They make neat compact clumps of grassy foliage, which serve well, as Gerard says, "for the bordering up of beds and bankes." Four forms of the native thrift; **A. *m. Splendens*,** which has larger more intense carmine flowers than the common wild form, **A. *m. Nifty Thrifty*,** which has pink flowers and cream splashed foliage, **A. *m. Rubrifolia*,** which has deep bronze foliage and deep pink flowers and **A. *m. Alba*,** a rare white form, native to our rocks and shores. All forms flower May-July and grow to about 15cm. tall and 25cm. in spread. Armerias make lovely cushion plants in paving or the gravel garden. Used in Tudor period in the making of knot gardens. The Rev William Hanbury writing in 1770 says "When they are in blow, and the first flowers decay, their stalks should be cut up to the bottom with scissors or the like, otherwise their withered dead heads and stalks will much injure the beauty of those in bloom, and have a very dull, indolent and disagreeable look." **Armeria pseudoarmeria** is slightly taller, has larger flowerheads and neater foliage than A. maritima. Its flowers are a delicate mauve. Looks lovely with the white perennial stock Matthiola fruticulosa Alba.

ARTEMISIA. (Asteraceae). Artemisia is the Greek word for wormwood. With the exception of **Artemisia *lactiflora Guizhou*,** all the varieties are aromatic and exceptionally drought tolerant. **A. *Limelight*** is a bold variegated form with deeply cut, cream splashed, deep green leaves. Very useful for early bright colour in March. **A. *Lambrook Mist*** forms a small woody shrub with deeply cut silver leaves Ht. 85cm. Sp. 45cm. Unlike A. Powys Castle, it does not resent a brutal cut back in autumn.

Aquilegia canadensis *(3)*

G Stuart Thomas suggests using A. Lambrook Mist as a backdrop for Nerine Bowdenii, to which I would add a carpet of Geranium Dusky Crug. The species plant **A. ludoviciana** in its original species form always looks to me underengineered for its Ht. 90cm.; **A. *l. Valerie Finnis*** carries her 40cm. of highly cut silver foliage with much more assurance. **A. *pontica*** has grey, fine silver foliage and grows to a modest 60cm. It has a vigorous, spreading habit. **A. *schmidtiana Nana*** is a low growing form with very fine, intensely silver silky foliage. Ht. 10cm. Sp. 25cm. **A. *lactiflora Guizhou*** is an altogether different beast. It has erect mahogany brown stems, black green foliage and bears creamy white flowers in August. It grows to about 125cm. in height, Sp. 80cm. and is a wonderful upstanding plant, which needs no staking. It is happy in clay and unlike the silver leafed artemisias needs plentiful moisture. Looks good with Althaea officinalis and Ligularias.

ARTHROPODIUM. (Asparagaceae). **A. *cirrhatum*** has enormous panicles of white flowers and bold strap-like light green leaves. A native of New Zealand, in a conservatory it will, with sufficient feeding with tomato fertiliser, flower throughout the year. Hardy in sheltered gardens. Beware of slugs! A plant of enormous grace and elegance. Ht. 60cm. Sp. 70cm. Its diminutive sister **A. *candidum Purpureum*** makes small, grassy clumps of deep bronze foliage with starry white flowers. Good in paving or gravel. June-July. Ht. 25cm. Sp. 30cm. Self-seeds.

ASCLEPIA. (Asclepiadaceae). Asclepias which come from North America get their English name Silkweed from their silky seeds. All asclepias need full sun. **A. *syriaca*** bears heads of scented pink flowers in June-Aug and grows to a mighty 150cm. tall. Sp. 50cm. It needs a reasonably fertile soil. **A. *tuberosa*** needs a hot position and dry soil. It bears brilliant orange flowers in July-Sept, which look wonderful with Convolvulus mauretanicus. Ht. 50cm. Sp. 25cm.

ASPHODELINE. (Asphodelaceae). ***A. liburnica*** bears delicious pale yellow, green striped flowers from June to August. It has very fine needle like glaucous foliage and a gentle willowy habit. Ht. 90cm. Sp. 30cm. A. liburnica is longer flowering and much more elegant than the more common **A. *lutea*** which bears loud chrome yellow flowers in May-June. A plant without subtlety - its mature flowering stems, before the flowers open, are a brazen homage to priapic vanity. Ht. 100cm. Sp. 30cm. The best thing about Asphodeline lutea is definitely its bold evergreen rosettes of glaucous leaves which brighten up the border in winter and make a wonderful backdrop for snowdrops. Its shiny black seed heads are not bad either. Both species of Asphodeline are native to the Mediterranean and thrive in poor dry soil, storing moisture in their fleshy roots.

ASTELIA. (Asteliaceae). **A. *chathamica Silver Spear*** is one of my two favourite plants for large pots and tubs. Its glistening silver, Phormium-like leaves make bold architectural clumps up to 150cm. tall with a similar spread. A native to New Zealand it is hardy here in our well drained soil. Should be fed and watered liberally during the growing season. I got spectacular growth on a plant one year by standing it all summer in a bowl of water!

ASTER. (Asteraceae). Asters derive their name from the Latin for a star. They are members of the daisy family and like full sun and well drained, moisture-retentive soil. All the following asters are mildew-free apart from the novi-belgii hybrids. **A. *amellus*** is a native of Italy and was introduced into this country in the 16th Century. All varieties of A. amellus flower Aug-Sept and should be allowed a spread of about 40cm. **A. *a. King George*** has showy lavender blue flowers. Ht. 60cm. Christopher Lloyd

Argemone mexicana *(3)*

recommends planting it with Nerine bowdenii. **A. *a. Pink* Zenith** has good clear pink flowers. Ht. 60cm. Sp. 50cm. **A. x *cordifolius Little Carlow*** bears bright lavender blue (2.5cm) flowers in great profusion in Sept-Oct. Ht. 120cm. Sp. 50cm. **A. *divaricatus***, a native of the U.S.A., has handsome foliage for an aster – small, dark, bronze green leaves - and clouds of tiny white flowers. June-Sept. Ht. 60cm. Sp. 50cm. It has a somewhat lax habit and is suitable for growing on banks or over walls. G. S. Thomas recommends it for draping over spent Bergenias in late summer, like a wedding veil on a pig some might say. **A. *dumosus Blue Saphire*.** Blue flowers June-Oct. A greatly improved variety. Ht. 35cm. Sp. 30cm. **A. *ericoides. Erlkonig*** has narrow green foliage and clouds of small blue flowers. Ht. 120cm. **A. *e. Pink Cloud*** is similar but shorter at 90cm. **A. *e. Ruth McConnell*** has lilac pink flowers with a dark disc. Ht. 90cm. All varieties of A. ericoides quickly make large clumps and flower Sept-Oct. **A. *x frikartii Monch*** has lavender blue flowers and I used to think was the best of all asters. In flower from July to November. Non invasive. It likes rich moist soil and dislikes being overcrowded. Ht. 75cm. Sp. 30cm. **A. *x frikartii Jungfrau*** now has superceeded A. f. Monch in my affections! It has truer blue flowers and a a more regular upright habit. **A. *lateriflorus Coombe Fishacre*.** Purple pink flowers with yellow centres, which quickly turn rosy brown. One of the most colourful of the small flowered asters. Sept-Oct. Ht. 60cm. Sp. 30cm. **A. *l. Prince*** makes an upright branching plant, well clad in dark, purple-bronze foliage. It needs no staking. It has small, white flowers with mauve pompon centres. Distinguished from A. l. horizontalis by its purple stems, its leaves retaining their dark colouring throughout the season and by flowering earlier – from the beginning of October through into late November. In my frosty garden the flowers of the later flowering A. l. horizontalis were always snubbed by frost when they scarce had opened their buds. After flowering A. l. Prince is still decorative, making a valuable vehicle for the frost. A plant in which the flowers and the foliage are in perfect balance. Shrubby looking, it makes a good alternative to lavender as a small hedge bordering a path. Should be cut to the ground in March. Ht. 90cm. Sp. 60cm.

The **Aster novae-angliae hybrids** unlike the novi-belgii hybrids are both mildew free and self supporting. They all flower from August to October and make up quickly into strong clumps – perhaps 45cm. in spread. **A. *n. a. Andenken an Alma Potschke*** has vivid carmine magenta flowers. Graham Stuart Thomas rightly says it "has enough colour in its glowing cerise scarlet to warm the most drear autumn day." Ht. 100cm. **A. *n. a. Harrington Pink*** has soft salmon pink flowers. Ht. 125cm. **A. *n. a. Purple Dome*** has purple flowers. Ht. 60cm. **A. *n. a. Septemberrubin*** has deep rose purple flowers. Ht. 100cm.

The **Aster novi-belgii hybrids** provide strong colour in September-October but are not mildew free, though plants grown in rich moist soil and full sun are seldom noticeably affected. All varieties spread rapidly and after two years are likely to make clumps at least 60 cm. in diameter. They can be dug up and split in both autumn or spring. **A. *n. b. Fellowship*** has sensuous, pale pink, double flowers. According to Bob Brown this plant has more mildew resistance than other n. belgii hybrids. Ht. 95cm. **A. *n. b. Litte Boy Blue*.** Useful shorter variety with dark blue flowers. Ht. 60cm. **A. *n. b. Little Boy Red*.** Compact variety. Rose red flowers **A. *n. b. Marie Ballard*** has full double, blue flowers. Ht. 90cm. **A. *n. b. Porzellan*** has delightful, pale, china blue flowers. Ht. 90cm. **A. *n. b. Twilight*** has violet blue flowers. Ht. 70cm. **A. *n. b. Winston Churchill*** has heads of bright single carmine flowers. Ht. 80cm.

Asclepia tuberosa *(3)*

A. ***tongolensis Berggarten*** is a completely different creature from all the above. A demure alpine plant from Western China, it makes neat basal clumps of hairy dark green leaves. Large lavender flowers with orange centres are borne on 60cm. stems in May-June.

ASTRANTIA. (Apiaceae). Native to Switzerland. The Rev William Hanbury wrote in 1770 that Astrantias "being flowers of no great beauty, the very worst part of the garden should be assigned to them." How tastes change! **A. *major Claret.*** I include this wonderful selection of Piet Oudolf's here, rather than in the shade section as for the best, richest, deepest, darkest red flowers, it needs full sun. In shade the flowers are still beautiful but somewhat paler. June-Oct. Ht. 70cm. Sp. 25cm. Sumptuous. **A. *m. Hadspen's Blood*** is a weak plant; A. m. Claret is a much stronger grower and has flowers which are just as rich in colour; the only visible difference between the two varieties is that A. m. Claret is 15cm. taller. **A. *m. Abbey Road*** is a newer and taller still variety with the darkest purple flowers and striking purple stems. Ht. 85cm. Sp. 25cm. **A. *m. Sunningdale*** like the dark flowered varieties needs to be grown in full sun, but for the benefit of its foliage, not its flowers. Dramatic cream splashed foliage in spring. The ordinary pinkish white flowers June-November. Ht. 55cm. Sp. 25cm.

AUBRIETIA. (Brassicaceae). "It is indeed a wonder with how greedy a zeal this Levantine race, from Lebanon and Olympus and Asia, has adopted the English climate" (Reginald Farrer). As Farrer goes on to say Aubrietia in fact positively dislikes torrid heat. Aubrietias look wonderful growing in stone walls and are great limelovers. **A. *Double Stockflowered Pink*** has large double light pink flowers. **A. *Elstead Giant Purple*** has very large dark purple flowers. **A. *Gloriosa*** has pink flowers. **A. *Red Carpet*** has large deep red flowers. **A. *Sauerland*** has light blue flowers.

BAPTISIA. (Leguminosae). **B. *australis.*** The name Baptisia is derived from the Greek for to dye, an indigo substitute being obtained from the plant. Baptisia australis is a member of the pea family and comes from the U. S. A. In the wild it grows on moist river banks. It is an extremely hardy, long lived plant. In July-August, it bears an abundance of indigo blue 'pea' flowers. It grows to 140cm. in height and makes a spread of about 70cm. A mature plant is a spectacular sight! Needs rich, moisture retentive acid or neutral soil. Dislikes being crowded.

BELAMCANDA. (Iridaceae) **syn. Iris domestica. B. *chinensis*.** Irregularly branched stems display yellow flowers brightly spotted with orange. June-July. Rich green sword shaped leaves. Seed pods split open in autumn to reveal decorative shiny, black seeds. Ht. 90cm. Sp. 30cm. Well drained fertile soil. Mulch in winter.

BERKHEYA. (Asteraceae). **B. *purpurea*** bears huge, (10cm) black centred, lavender daisy flowers in clusters from June to Nov. Sun. Prickly greyish green foliage. Easy. A must for the gravel garden. Well drained soil. S. African but hardy. Ht. 60cm. Sp. 40cm.

BESCHORNERIA. (Asparagaceae). Pink and green flowers. Electric pink stems. Flowers biennially. Evergreen rosette. Ht. 100cm. Sp. 50cm. Hardy. Collected from a high altitude site in Northern Mexico.

BUDDLEJA. (Scrophulariaceae). Buddlejas grow well on our light sandy soil. Like Lavateras they are good for a quick effect. **B. *Black Knight*** is a wonderful dark royal blue. Ht. 200cm. Sp. 120cm. Prune hard in spring.

Asphodelus luteus *(3)*

BUPLEURUM. (Apiaceae). **B. *fruticosum*.** Leaden green leaves. Umbels of small greenish yellow flowers. July-Sept. Ht. 200cm. Sp. 200cm. Sun. Moist fertile soil. Technically a shrub, takes several years to become properly established. Gives a lovely yellowy green effect, similar to a Euphorbia but to my mind more elegant. Lovely with Verbena bonariensis.

CALAMINTHA. (Lamiaceae). Mediterranean plant. **C. *nepeta Blue Cloud.*** Clouds of blue flowers on small wiry plants, covered in tiny, mint scented foliage. Frequented by bees. Good in the gravel garden. Ht. 35cm. Sp. 40cm. Even prettier to my mind is **C. *n. Alba,*** whose dainty white flowers are a welcome antidote to an overdose of helianthemums. Both Calaminthas flower throughout the summer from July to September.

CALTHA. (Ranunculaceae). The word caltha derives from the Greek 'calathus', a cup. **C. *palustris*** has grown in England since before the Ice Age and has attracted masses of vernacular names most of which are hard to explain. Horse Blobs? Water Dragons? Drunkards? The wild single Caltha is one of the joys of spring; in greyest March its buttery orange flowers light up the desolate muddy verges of rivers and ponds. Its foliage, as Gerard says, is a "galant green". In Lapland where it blooms at the end of May, it is regarded as a harbinger of spring. In England where it starts blooming at the end of April it was, according to Hanbury, "usually gathered for strewing before the doors on the eve of May-day, to usher in that welcome month." **C. *p. Flore Pleno*** has glorious brazen, yellow, fully double flowers. Utterly tasteless and wonderful! April-May. Ht. 30cm. Sp. 40cm. **C. *p. Alba*** is one of prettiest and most modest of plants. Stemless white flowers, resembling nothing so much as a clutch of bird's eggs (Beth Chatto's image, not mine) open out into shallow saucer shaped flowers embellished with central golden bosses. Although the main flush of flowers is in spring, a well established plant will continue throwing out odd flowers throughout the summer. C. Lloyd complains that the flowers are thin textured, but I think their delicate vulnerability is part of their charm. Rather Caltha palustris Alba any day than the heavy textured flowers of water lilies. Ht. 15cm. Sp. 30cm.

CAMASSIA. (Liliaceae). Natives of N. America. Introduced to England in 1827. In the wild flowers "in some places so profusely as to resemble lakes of blue water", (Alice Coats). The bulbs were apparently valued by the native North American Indians (and grizzly bears!) as a food stuff. Camassias thrive in rich moist soil and an open position. **C. *cusickii*** has ice blue flowers in June. Ht. 80cm. Sp. 30cm. **C. *leichtlinii caerulea*** bears deep blue flowers. May-June. Ht. 90cm. Sp. 30cm. **C. *l. alba*** is the very scarce white form. Lovely, ivory white, starry flowers in May-June. Ht. 90cm. Sp. 30cm. All varieties are easy to naturalise in a damp marshy meadow.

CAMPANULAS. (Campanulaceae). I have divided Campanulas into three groups: border plants for moisture retentive soil, border plants which like a drier soil and Alpine varieties suitable for planting in paving, walls and rockeries.

Moisture Loving Campanulas. C. *takesimana Elizabeth*, like C. Kent Belle, originates from Elizabeth Strangman's wonderful nursery in Kent, now, alas, no more. It has large, white, tubular bells heavily spotted pink. May-Aug. Ht. 50cm. Sp. 20cm. **C. *Kent Belle*,** is a chance hybrid from C. takesimana. It has glossy leaves and rich violet tubular flowers. In moist fertile soil grows to 75cm. Sp. 35cm. Shade tolerant. Repeat flowers if cut back after first flowering. June-Sept. **C. *Sarastro*** has fragrant dark purple flowers. Very free flowering. May-Sept. Ht. 60cm. Sp. 30cm. Similar to C. Kent Belle

Aster amellus *(3)*

but much shorter, which can be useful. Bob Brown rates Sarastro higher than Kent Belle, but for me the matt violet of Sarastro is dull compared to the beautiful deep purple gloss of Kent Belle. If you need something for the back of the border, I recommend **C. *lactiflora*. C. *l. Loddon Anna*** has white-throated shell pink flowers. June-Sept. Ht. 120cm. Sp. 60cm. The quite outstanding violet blue **C. *l. Pritchard's Variety*** has deep violet, white throated flowers. June-Sept. Ht. 130cm. Sp. 60cm. There is also a dwarf form called **C. *l. White Pouffe***, which makes neat mounds at the front of the border. Flowers June-Aug. Ht. 30cm. Sp. 30cm. **C. *latifolia*** is native to Northern Europe. It is a plant to be valued for its strong vertical lines and the great elegance of its pendant bellflowers. **C. *l. Alba*** has pure white flowers, Ht. 80cm, and **C. *l. Macrantha*** has purple bells, Ht. 100cm. Both will spread strongly in fertile soils, making large clumps in a couple of years. Sp. 80cm. They flower June-August. **C. *latiloba*** is a must for heavy soils. All varieties bear 90cm. spikes of large, open, saucer shaped flowers in June-July. Sp. 50cm. **C. *l. Highcliffe*** has medium blue flowers. **C. *l. Hidcote Amethyst*** has astonishing dusky amethyst flowers.

Campanulas for Drier Soils. C. *alliariifolia* has a rosette of heart shaped, mid green leaves from which it throws up sprays of ivory white flowers. June-August. Handsome mid-green heart shaped leaves. Rust and mildew free. Ht. 45cm. Sp. 30cm. **C. *persicifolia*** will grow in any well drained soil in sun or shade and flowers from June till late August. All varieties of C. persicifolia are delicate, modest plants and associate well with old fashioned roses and lilies. Each plant should be allowed a space about 25cm. in diameter. **C. *p. Alba*** is the pure white form. Ht. 80cm. **C. *p. Caerulea*** has large, single blue flowers. Ht. 80cm. **C. *p. Chettle Charm*** has large, pure white flowers, flushed with blue at the trumpets edge. Ht. 50cm. **C. *p. Gawen*** has semi-double creamy white flowers. Ht. 60cm. **C. *p. Kelly's Gold*** has brilliant gold foliage and white flowers. Ht. 60cm. **C. *p. Wortham Belle*** has light blue double flowers and is a strong grower with upright stems. Ht. 75cm. **C. *rapunculoides*** is "the most insatiable and irrepressible of beautiful weeds. If once its tall and arching stems of violet blue flowers prevail on you to admit it to your garden, neither you nor its choice inmates will ever know peace again". Thus wrote Reginald Farrer and of course he was right, C. rapunculoides has no place in the garden, but in the wild garden or orchard it is a splendid thing, well able to compete with grass. June-Aug. Ht. 80cm. Sp. 40cm. **C. *punctata*** is native to Japan and makes a useful groundcover plant for well drained soil. In heavier soil it tends to die out. All varieties have handsome rich dark green foliage. **C. *punctata Hotlips*** has tubular pale pink flowers, heavily spotted inside the tube with purple spots. Ht. 25cm. Sp. 30cm. May-July. **C. *p. Pantaloons***. Light purple, double flowers on strong upright stems. Good cut flower. June-July. Ht. 40cm. Sp. 40cm. **C. *p. Wedding Bells*** has double white flowers with pink flushes. July-Sept. Ht. 50cm. Sp. 30cm.

Alpine Campanulas. All these need full sun and well drained soil. **C. *Birch's Variety*** is a cross between C. portenschlagiana and C. poscharskyana. It has rich purple blue flowers. June-August. Good in gravel or walls. Ht. 15cm. Sp. 30cm. **C. *carpatica Alba***. Pure white flowers June-Aug. Compact plant. Good as edging. Ht. 25cm. Sp. 25cm. **C. *cochlearifolia* syn. Pusilla** is the daintiest of creatures. It grows only 10cm. tall, and is covered in the prettiest of tiny light blue bellflowers from June to August. It looks good in paving or a trough, where it will make a small mat perhaps 20cm. in diameter. The enchanting white form **C. *c. Alba*** looks stunning planted with Arthropodium candidum Purpureum, and **C. *c. Elizabeth Oliver***, an exquisite fully double blue. **C. *garganica*** is

Aster frikartii Monch *(2)*

another dwarf campanula but with much larger darker blue flowers from March till May. Good in gravel or paving. Ht. 10cm. Sp. 30cm. Sun and good drainage. Looks good in gravel and is normally evergreen. **C. *x haylodgensis*.** An enchanting miniature with double, lavender blue flowers. July-Sept. Ht. 15cm. Sp. 30cm. **C. *poscharskyana*** is native to Dalmatica. It is similar to C. garganica but a much more vigorous grower. Three varieties: **C. *p. Stella*** is a selection with particularly bright blue flowers, **C. *p. E. K. Frost*** has milky white flowers and **C. *p. Lisduggan*** has pink flowers and is less vigorous than the blue and white forms. All forms have a height of 25cm. and flower June-July. For a small front of border Campanula, which thrives in the driest, poor soil, plant the harebell, **C. *rotundifolia***, a native heath land plant. Endless, dainty blue flowers on 40cm. stems. Sp. 25cm. An ethereal delight May-Sept. Sun. Well drained soil. Looks good in paving or gravel with Erigeron karvinskianus and Convolvulus mauretanicus. **C. *Yvonne*.** Low spreading plant. Masses of upturned lilac-blue bell-like flowers. June-August. Ht. 10cm. Sp. 45cm.

CANNA. (Cannaceae). "Decried as vulgar by snooty private gardeners of the silver and grey, impeccable taste school." (Christopher Lloyd). **C. *Black Knight*** has the blackest leaves of all the cannas. Large red flowers. Ht. 120cm. **C. *City of Portland*** has pink flowers. 100cm. **C. *Durban*** has purple maroon leaves with red and yellow stripes. Shocking orange flowers. Ht. 150cm. **C. *Firebird*** has deep red flowers. Ht. 100cm. **C. *Golden Girl*** has yellow flowers covered in red spots. Ht. 60cm. **C. *Lucifer*** has red flowers edged with yellow. Ht. 60cm. **C. *Pink Sunburst*.** Maroon pink and green striped leaves. Soft salmon pink flowers. Aug-Oct. Compact. Ht. 50cm. Sp. 45cm. **C. *Pretoria* syn. C. Striped Beauty.** This is one of the key plants in the tropical garden at Great Dixter. Huge, pale yellow and green striped leaves. Large scarlet flowers. Sept-Oct. Ht. 120cm. **C. *Salmon Pink*** has pale pink flowers and yellowish foliage. Ht. 100cm. **C. *Stuttgart*** has orange flowers and green leaves with bold white variegation. Ht. 150cm. All Cannas flower from July onwards and in a single season make plants with a spread of about 35cm.Although Cannas are sufficiently hardy to overwinter in some warm London gardens, they are always best lifted, as they need forcing into growth in a greenhouse in the spring, if they are to make really impressive plants before autumn. The lifted tubers should be stored in peat and kept at a temperature of not less than ten degrees during winter. The tubers must not completely dry out, and should occasionally be slightly moistened. Repotted in April, they will be ready for planting out in June as established plants. Beware! Slugs love Cannas. 100cm.

CARYOPTERIS. (Lamiaceae). **C. *clandonensis*.** An attractive, silver leafed shrub from Eastern Asia, its name translates from the Greek as winged nut, a reference to the shape of the seed head. Caryopteris clandonensis, given good drainage, is stone hardy and a reliable provider of colour in the late summer garden. It makes a small shrub about 90cm. in diameter. To ensure a compact plant prune hard in spring. **C. *c. Worcester Gold*** has soft gold foliage and mid blue flowers. Aug-Sept and **C. *c. Kew Blue*** has plain grey green leaves and strong blue flowers. Both varieties flower Aug-Sept.

CATANACHE. (Asteraceae). The species plant was introduced from Italy by Gerard in 1597. The white flowered form was selected by a Mr Smith of Worcester in the early 19th Century. If, like me, you are a cornflower fan, Catanaches bridge the gap between the spring flowering Centaurea montanas and the late summer flowering Stokesias. One of the great beauties of this plant is its silver calyxes. Catanache is an essential plant for the

Auricula Arundel Stripe *(2)*

hot dry border. For a light airy effect, plant with Gaura lindheimeri. Plant in groups of at least three plants. Lovely as a cut flower. C. ***caerulea Major*** has large, rich blue flowers with black centres June-Sept and **C. *c. Alba,*** the white flowered form July-Sept. Both grow 70cm. tall and have a spread of 30 cm. Christopher Lloyd says of Catanache: "a mildly pleasing but second rate plant". So speaks one wont to glut himself on Dahlias and Cannas!

CENTAUREA. (Compositae). The name derives from Chiron the Centaur, who is supposed to have cured with Centaurea a poisoned wound given him by Hercules. All members of the knapweed family thrive in any well-drained, not bone-dry soil and full sun. C. ***atropurpurea***. Species plant from the Carpathian Mountains.Tight flower heads of a deep darkest crimson. August. Ht. 180cm. **C. *bella*** has pink flowers June-Aug and pristine silver foliage. Ht. 20cm. Sp. 30cm. Lovely in gravel. **C. *hypoleuca John Coutts*** has large, fragrant, deep rose flowers and deeply cut silver foliage. June-Aug. Ht. 50cm. Sp. 50cm. C. ***macrocephala***, native to the Caucasus, is a bold plant both in foliage and flower - sinew for borders of floppy prettiness. Huge, real gold flower heads open yellow. June-Aug. Large, heart shaped green leaves. Ht. 120cm. Sp. 30cm. **C. *montana***, a European native, was introduced into England before 1596. One of our oldest garden flowers, Gerard knew C. montana under the name of Blew Bottle. It is one of the great joys of spring. Its rich blue flowers, May-July, are offset by grey green leaves. Ht. 60cm., Sp. 40cm., It makes a lovely accompaniment for Lupins. A very choice, white form C. ***m. alba***, May-July, which looks very crisp with Nepeta Walkers Low. **C. *dealbata Steenbergii*** has finely divided, light green foliage and dark, carmine lilac flowers June-Aug. A robust plant. Ht. 75cm. Sp. 50cm.

CENTRANTHUS. (Asteraceae). The Greek name refers to the spurs on the Valerian's flowers. A native to Southern Europe, it appears to have been introduced to England in the 17th Century. Gerard considered it a great ornament to his garden and described it as not common in England. It appears to have been introduced into England for purely decorative reasons although, in France and Italy the very young leaves are added to a green salad. Its seed was formerly used in embalming. Centranthus is a wonderful plant for dry places. It never looks handsomer than when growing out of an old stone wall, effortlessly thriving on a diet of dry lime mortar, always to be clothed in the freshest of light green foliage. The species, **C. *ruber*** has flowers of a rusty red. June-September. Ht. 75cm. Sp. 40cm. It looks lovely with yellow Euphorbias. **C. *r. Alba***, the white form flowers June-August. Ht. 75cm. Sp. 40cm. It looks well in gravel with Centaurea bella. Centranthus should be cut down to the ground twice annually, once in February, removing all the previous season's manky growth, and again in early summer, to encourage a second flowering. This plant will grow anywhere, given sun and good drainage. Centranthus should not be confused with the true Valerian; see Valeriana officinalis.

CEPHALARIA. (Caprifoliaceae). The name Cephalaria derives from the Greek for head, the flowers being borne in round heads. Cephalarias are closely related to the Scabious family. They need moisture-retentive soil and full sun. Both C. alpina and C. gigantea flower June-Aug. C. ***alpina*** grows 125cm. high, spread 60cm. and is like a more compact, paler flowered form of C. Gigantea. C. ***gigantea*** makes a magnificent, architectural specimen, growing to 200cm. Sp. 90cm. It has the softest of yellow flowers, borne on widespreading branches with bold, rich green, pinnate foliage. June-August. C. ***dipsacoides*** grows as tall as C. gigantea but has smaller and more numerous greenish yellow flowers. Cephalarias tolerate heavy, badly drained clay.

Auricula Blue Nile *(2)*

CERASTIUM. (Caryophyllaceae). C. ***biebersteinii*** is less invasive than the common C. tomentosum. Its foliage is also more intensely silver and it flowers over a longer period. White flowers. May-July. Ht. 15cm. Sp. 30cm. Well drained soil. Useful on hot banks and in walls and paving. Reginald Farrer rates C. biebersteinii as a peach among cerastiums.
CERATOSTIGMA. (Plumbaginaceae). C. ***willmottianum*** is a Chinese member of the plumbago family. Ceratostigma means horned stigma. A compact shrub, covered in gentian blue flowers Aug-Oct. Full sun and well drained soil. Lovely with Salvia microphylla. Ht. 75cm. Sp. 30cm. Christopher Lloyd suggests planting with red fuchsias. C. ***plumbaginoides***. A creeping, entirely herbaceous form which is good for weaving in and out of the front of the border. Gentian blue flowers. Good autumn leaf colour. Ht. 30cm. Sp. 45cm.
CERINTHE. (Boraginaceae). C. ***major Purpurascens***. Blue foliage and rich purple flowers. This delightful biennial is a native of the Mediterranean, where it grows wild in meadows with oxeye daisies. Although known to English botanists at least since the beginning of the 19th Century (see our illustration from The New Botanic Garden published in 1812) Cerinthe has only recently become a fashionable plant in this country. Flowers April-June. Ht. 20cm. Sp. 35cm. An easy plant for well drained soil; once established, it self seeds freely.
CHAMAEMELUM. (Asteraceae). C. ***nobile Flore Pleno***. 'Chamai' is the Greek for on the ground i.e. prostrate, and melon means apple, which is supposed to refer to its scent. Although as Geoffrey Grigson says the wild chamomile Matricaria chamomilla has far more scent of apple. Chamaemelum used to be called Anthemis and is the Chamomile of herb teas, the commercial production of which this beautiful, double flowered plant is commonly used. The single flowered form is a dreary thing with its draggled, greyish white flowers. The double has elegant creamy white pompons. Lovely for cutting in mixed bunches, particularly with roses. And of course it can be used either fresh or dried as an herbal tea. Gerard describes it as having "a force to digest, slacken, and rarefie....wherefore it is a speciall helpe against wearisomnesse, it easeth and mitigateth paine, it mollifieth and suppleth." It was used in the Tudor period as a strewing herb. In Spain it is still used today, to provide the characteristic flavour to Manzanilla. Has lovely, fresh, bright green, winter foliage. Flowers July-Aug. Ht 30cm. Sp. 40cm. Bright evergreen ferny winter foliage. C. ***n. Treneague***. The flowerless lawn chamomile, ideal for a small aromatic lawn. Looks brighter and more cheerful than grass in winter. Self maintaining apart from an occasional trim with shears or a flymo. There is a tradition that Drake played his famous game of bowls on a chamomile lawn.
CHELONE. (Plantaginaceae). C. ***obliqua***. The common name for this plant is Turtle Head, a name which derives from the curious shape of the pink flowers, which resemble the open mouths of turtles. Pink turtles! Lord Berners! August-September. Moist soil. Upright 60cm. stems. The very beautiful C. ***o. Alba*** has pale ivory flowers. Both varieties make excellent cut flowers. Slugs must be controlled or they will gluttonise on the foliage.
CHRYSANTHEMUM. (Asteraceae). 'Chrysos' is Greek for gold, 'anthos' for flower. Chrysanthemums have been cherished in China for 500 years. Why are people snobbish about hardy chrysanthemums? For me they are essential, cottage garden plants, filling the garden with colour from October through to December. Perhaps it is the association with florists' Chrysanthemums, those impossibly heavy headed monsters in hard white

Auricula Trudy *(2)*

or searing yellow, that makes people disdain their lovely garden counterparts. Hardy Chrysanthemums are easy to grow. They like rich soil and reasonable moisture. They must have full sun and are best not planted in frost pockets, as this will drastically truncate their flowering season in a cold autumn. Chrysanthemums clump up readily and should be lifted and divided in early spring. To encourage more compact plants, better clothed with foliage, I cut all my Chrysanthemums back hard in early June. At the same time I remove all weak and spindly canes. This does not delay their flowering and results in much sturdier plants. Of all the hardy Chrysanthemums I think my favourite to be **C. *The Emperor of China*.** This very lovely old variety (some, on the basis of painted scrolls, believe it to be of ancient Chinese origin) has the softest of rose pink double flowers. The petals are somewhat quilled, and this gives the flowers great distinction and character. An added bonus is that the foliage becomes richly suffused and veined with crimson in autumn. There are also the following additional varieties: **C. *Cottage Apricot*** single warm apricot flowers, **C. *Duchess of Edinburgh*** semi double, deep red flowers, **C. *Clara Curtis*** single, shell pink flowers, **C. *Wedding Day*** single white flowers and **C. *Mary Stoker*** pale biscuit coloured, single flowers. All flower in October and can still be in flower in a sheltered spot in December. All are about 100cm. tall apart from the pompon varieties, which are about 80cm.

CICHORIUM. (Asteraceae). **C. *intybus*.** Member of the daisy family. The ever lovely wild chicory. A native to the Mediterranean, this is a plant which thrives in the most arid of soils, protected from drought by its deep taproot. Indomitable. Its foliage and stems skeletonised by sun and drought, it will still contrive to produce its limpid sky blue flowers. Beloved by mediaeval illuminators, a wild chicory flower will be found in the margins of almost any Book of Hours. Ht. 90cm. Sp. 20cm. Lovely with the yellow Tree Lupin. Two double forms: **C. *i. Flore Pleno Album*.** Pure white flowers July-Oct. Ht. 110cm. Sp. 20cm. and **C. *i. Flore Pleno Roseum*.** Rose pink flowers. Ht. 100cm. Sp. 20cm.

CIRSIUM. (Asteraceae). **C. *diacanthus* syn. Ptilostemon Afer.** Bold strong, white variegation on dark green leaves. Viciously sharp thorns! Purple flowers. July-Aug. Ht. 45cm. Sp. 30cm. Self seeds. **C. *rivulare Atropurpureum*.** Dark red thistle knobs held over a rosette of handsome cut green foliage. No prickles. Very extended flowering season. March-June and August-December. Ht. 100cm. Sp. 35cm. **C. *Mount Etna*.** Pink and white thistle flowers July-Sept. Large, mid green thistle leaves. No prickles. Ht. 60. Sp 30cm. Cirsiums need sun and will grow in any not too dry well drained soil. Bees and butterflies congregrate on their flower heads.

CISTUS. (Cistaceae). **C. *ladanifer*.** The lovely gum cistus. White flowers with maroon splodges around central boss of gold stamens. June-July. Sticky aromatic dark green leaves. Ht. 120cm. Sp. 120cm. **C *obtusifolius*** is a dwarf Cistus which grows to only 25cm. Sp. 60cm. In May it covers itself in white flowers. Matt green foliage. Rare thing for a Cistus, looks presentable even in the middle of winter. Cistuses need full sun and well drained soil.

CLEMATIS. (Ranunculaceae). Clematis gets its name from the Greek for tendril, 'clema'. Herbaceous varieties can be cut to the ground each winter and are excellent for scrambling through shrubs or for covering a bank. Once established they tolerate dry soil and will grow in sun or part shade. All the following varieties are highly scented. **C. *bonstedtii Crepuscule*** has sky blue flowers June-Sept. Ht. 80cm. Sp. 150cm. **C. *j. Mrs Robert Brydon*** has pale watery blue flowers. August-October. Ht. 60cm.

Auricula Spring Meadow *(2)*

Sp. 150cm. C. ***recta Foliis Purpureus*** has bronze foliage and milky blue flowers. June-July. Ht. 100cm. Sp. 150cm.
CLIVIA. (Amaryllidaceae). C. ***miniata***. Named after a certain 19th Century Duchess of Northumberland, whose maiden name was Clive, the Clivia is a native of Natal. My clone, C. ***m. Mrs. P. Loftus*** has narrower, more elegant foliage than the type. The flower trumpets too are narrower and a soft pastel orange, rather than the normal shrill orange. Easy. Dislikes direct sunlight. Must be kept frost-free. Tolerates gross neglect. When not in flower, attractive as a foliage plant. Usually January flowering. Ht. 60cm. Sp. 60cm.
CODONOPSIS. (Campanulaceae). C. ***clematidea***. 'Codonopsis' is the Greek for bellshaped and 'clematidea' refers to its clinging tendrils. A member of the Campanula family, it is a native of the Himalayas. C. clematidea has somewhat lax 60cm. stems, which support single, pale blue bells. The plant's true beauty is the interior of the flowers. The grey blue bells have dark stamens, the base of which are surrounded by a ring of vivid orange. June-July. Sp. 10cm.
COLCHICUM. (Colchicaceae). Colchicums get their name from the town of Colchis, where Medea, administering an elixir to Aeson to restore his youth, spilt some drops on the ground from which Colchicums sprang. The early herbalists named the colchicum The Son Before the Father because they believed it produced its seed before its flower, the flower appearing like a "posthumous poem." Colchicums have attracted a host of English names, (Colchicum autumnale is native to Britain), mostly referring to the flowers appearing before the leaves. Naked Boys in Somerset, Naked Ladies in Dorset. To each according to his or her taste! Although poisonous in the wrong dosage, colchicums are still used successfully in the treatment of gout. C. ***speciosum Album***. A rare plant of surpassing beauty. Huge white chalice shaped flowers in September. Ht. 20cm. Sp. 30cm. Any well drained soil.
COMMELINA. (Commelinaceae). C. ***coelestis***. Named in honour of the Dutch 17th Century botanist Jan Commelin, C. Coelestis was introduced into England from Mexico in 1737. In appearance somewhat like an exotic Tradescantia. But where Tradescantias have flowers of subdued tones, the Commelina positively zings. Its flowers are of a vibrant, azure blue. Aug-Sept. Ht. 45cm. Sp. 10cm. Despite its Mexican origins, C. coelestis, like Dahlias, is remarkably hardy in well-drained soil, but does not tolerate being waterlogged in winter. If you suffer from clay, you should dig up the Commelina tubers in late autumn and store them in peat for the winter. In normal soil, they will, like Dahlias, thrive without any such coddling. For those of a horticulturally nervous disposition, there is reassurance; even if your Commelina perishes, it will have seeded itself. An observant eye will be able to pick out its seedlings by their distinctive, rich green, lanceolate foliage. Seedlings appear in May or June and will flower the same summer. Commelinas suffer from no disease or parasite, save the dreaded slug, which gluttonously indulges itself on the fleshy leaves. Commelinas are plants for the morning, normally by early afternoon the flowers have closed, although on cloudy days the flowers remain open all day, as if to compensate us with their wonderful azure for the lack of zing in the sky. On a more mundane note the roots boiled and served with a white sauce are supposed to constitute "an agreeable table vegetable."
CONVOLVULUS. (Convolvulaceae). Two Mediterranean cousins of the common Bindweed. Both have the reputation of being one sock short of total hardiness, but seem to survive in all but the frostiest of gardens. C. ***althaeoides*** bears endless, dusky pink

Belamcanda chinensis *(3)*

flowers July-Oct. It has dramatic, silver filigreed foliage, which perfectly offsets the flowers. Can be invasive, so best confined to a pot. Looks wonderful in urns. Dislikes dry sandy soil. Ht. 25cm. Sp. 90cm. **C. *mauritanicus*** likes a lighter soil than C. althaeoides. It has silver green, uncut leaves and bears deep blue flowers throughout the summer. Ht. 25cm. Sp. 60cm. Plant it with E. karvinskianus. Lovely in pots and hanging baskets.

CORDYLINE. (Asparagaceae). Cordylines make graceful palm like trees up to 300cm. high. They are much hardier than generally supposed. If planted out of doors, it is a good idea to tie up the leaves of young plants in winter. This will protect the crown from winter wet. Architectural. **C. *australis. Red Star*** has the reddest foliage yet. A Cordyline makes an excellent sentinel planted in the front garden of an Edwardian or Victorian terrace house, guarding the approach to the stained glass door.

COREOPSIS. (Asteraceae). Coreopsis gets its name from the appearance of its fruit. Coreopsis translated literally means buglike. They are native to North America. All Coreopsis are sun lovers and thrive in any reasonably fertile, moisture-retentive soil. **C. *lanceolata Walter*** makes a very neat plant of fresh green foliage, covered from May to October with crisp yellow daisies with an inner brown ring. Clump forming. Ht. 20cm. Sp. 30cm. **C. *verticillata*** was introduced to England as an ornamental in 1759. In America it was used to dye cloth red. C. verticillata has very fine leafed, dark green foliage. It spreads stoloniferously. Two yellow varieties both flower June-Sept and are about 45cm. high with a spread of 30cm: **C. *v. Grandiflora*** has large golden yellow flowers and **C. *v. Moonbeam*** has creamy yellow flowers. **C. *rosea American Dream*** makes a dense forest of stems, tipped with rose pink flowers July-Aug and is similar in habit and foliage to C. verticillata. Ht. 30cm. Sp 30cm.

CORONILLA. (Papilionaceae). **C. *varia*.** Pale pink vetch-like flowers. May-June. Ht. 60cm. Spreading habit. Will grow in the poorest of dry soil. Looks good in gravel.

COSMOS. (Asteraceae). **C. *atrosanguineus*.** Deepest red, almost black, velvety flowers over pretty cut foliage. Aug-Sept. Half hardy. Ht. 60cm. Sp. 20cm. Needs rich feeding. Plant extra deep in well drained soil and protect with mulch in winter. Cosmos is another Mexican tuberous plant, like Commelina and Dahlias, which seems to be reasonably hardy, given good drainage. Perhaps best with only morning sun. Can wilt in intense heat.

CRAMBE. (Brassicaceae). Crambe is a member of the cabbage family; Crambe is in fact the Greek word for cabbage. Two varieties, **C. *cordifolia*** and **C. *maritima*.** C. cordifolia, a native of the Caucasus has large limp hairy green leaves surmounted in June-July by a vast cloud of Gypsophilla-like flowers - the cloud being up to 150cm. in diameter. The plant's total height is about 200cm. Its bold basal rosette reaches about 90cm. in diameter. After flowering, it is best to cut down both flowering stem and leaves, as "in their senile decay, ... they exhale an appalling odour of decayed Cabbage stalks". E. A. Bowles. A handsome rosette of fresh foliage is then produced. From mine own experience, reluctance to promptly remove the old flowering stem may result in the plant's failure to flower in the following year. This is a pity because, stench be damned, the weathered flower stems are magnificent frosted in winter. Crambe is generally free from pests, apart from the larvae of the Cabbage White butterfly, which only too frequently are to be found on the underside of the leaves, and will, as Bowles laconically remarks, skeletonise them. Bowles planted his Crambe cordifolia in a bed with Acanthus. Crambe cordifolia certainly needs architectural companions. Geranium

Brunnera macrophylla Langtrees *(2)*

palmatum might be another happy association. C. maritima or sea kale is a much more compact plant. Decorative waxy purple leaf buds open in spring to reveal rubbery blue leaves. This is one of the two best blue leafed plants! Honey scented white flowers. May-July Ht. 60cm. Sp. 50cm. A native of our seashore it lives "upon the bayche and brimmes of the sea, where there is no earth to be seen, but sande and rowling pebble stones." (Anon). Its seeds are distributed by the tide. Crambe maritima, blanched by earthing up - necessary to eliminate the bitterness - has been cherished as a vegetable since the Middle Ages and is indeed one of our very few native vegetables. Both Crambes are taprooted and when established will withstand great drought.

CROCOSMIAS. (Iridaceae). Their Greek name signifies crocus-smelling, which refers to the smell the dried flowers give off, when immersed in water. Crocosmias have a P.R. problem. Many gardeners still shudder at the name, instantly visualising that ubiquitous, dreary, orange weed, Crocosmia x crocosmiflora. Such prejudice needs to be overcome. Many Crocosmias have a very unmontbretia like intensity of colour. C. ***Amber Glow*** has orange red flowers with bronze foliage. July-Sept. Ht. 40cm. C. ***Babylon*** has large orange red flowers with brown markings. Aug-Sept. Ht.75cm. Sp. C. ***Carmine Brilliant*** has carmine red flowers July-Sept. Ht. 35cm. C. ***Dusky Maiden*** has brownish orange flowers Aug-Oct. Ht. 55cm. C. ***Emily Mckenzie*** is perhaps the latest to flower and has huge orange flowers, dramatically splashed with deepest red. Aug-Oct. Ht. 60cm. C. ***Firebrand*** has fiercesome heads of small ember red flowers brandished upwards. July-Sept. Ht. 35cm. C. ***Gerbe D'Or*** has brown leaves and apricot flowers. July-Sept. Similar to C. Solfataire but taller, hardier and more vigorous. Ht. 60cm. C. ***George Davidson*** has big, soft apricot flowers in large sprays with conspicuous green calyxes. Aug-Sept. Ht. 60cm. C. ***Honey Angels*** has dainty clear yellow flowers. July-Sept. Ht. 35cm. C. ***Lucifer*** has the purest, fiery scarlet flowers. It is early flowering and by far the tallest variety. It also has uniquely wonderful architectural foliage. Best planted in big clumps. June-Aug. Ht. 100cm. C. ***Mistral*** has upturned orange flowers, which turn pinkish. Jul-Sep. Pleated foliage Ht. 100cm. C. ***Saracen*** has sprays of large tomato red flowers July-Sept. A useful free flowering compact variety. Ht. 55cm. C. ***Short Apricot*** profuse, apricot yellow flowers. July-Sept. Ht. 45cm. Sp. 30cm. C. ***Solfataire*** apricot yellow flowers with bronze foliage. July-Sept. Ht. 45cm. Sp. 30cm. C. ***Star Of The East*** huge, (up to 5cm. across) pale orange, star shaped flowers. July-Oct. Ht. 60cm. Sp. 30cm. C. ***Zeal Tan*** a useful short variety with deep red flowers. July-Sept. Ht. 40cm. Sp. 40cm. Crocosmias are on the whole unfussy, sunloving plants, thriving in more or less any soil, including heavy-duty clay. The exceptions to this are Solfataire and Star of the East, both of which need sharp drainage to safely overwinter. Crocosmias do sometimes take a year or so to settle down after planting and can be shy in flowering the first year - what Christopher Lloyd calls "the debutante syndrome." Crocosmias look tasteful with the strong blues of Salvia patens and S. guarantica. If you are tasteless like me, plant them with red Dahlias, and Helenium Moerheim Beauty.

CYNARA. (Asteraceae). C. ***cardunculus.*** The Cardoon gets its Greek name from the resemblance of its phyllaries to dogs teeth. In the Victorian period it was grown as a vegetable, the young leaves were blanched like celery and eaten boiled. Whatever its merits as a vegetable, it makes a superb architectural plant with its magnificent, deeply cut silver foliage and boldly brandished light violet flowers. Aug-Sept. In winter the dead flowering stems continue to provide good vertical lines. The Cardoon is happy in

Bupleurum fruticosum *(3)*

either acid or limey soil. Being taprooted, it will withstand drought when established, but is more magnificent, when grown in soil with some moisture. Immaculately evergreen in mild winters. Flowering stems: 200cm. 100cm. mounds of foliage. Full sun. Plant with Melianthus major, Acanthus mollis, Inula magnifica, Geranium palmatum and Crambe cordifolia. **C. *scolymus*.** The Globe artichoke is similar to the above but with less prickly flowers and greener foliage.

CYNOGLOSSUM (Boraginaceae). **C. *nervosum*** has bright deep blue flowers in May-June. Does well on poor soil. Christopher Lloyd suggests planting with Cerastium tomentosum and yellow lupins. Ht. 60cm. Sp. 40cm.

CYRTANTHUS. (Amaryllidaceae). **C. *purpureus.*** Bears the English name Scarborough Lily, because, they were discovered growing on Scarborough beach, having been washed overboard from a ship. Cyrtanthus translates as curved flower which refers to the Vallota's curved perianth tube. Vallotas are native to S. Africa. They are hardy in Cornwall and Donegal where I have seen them growing happily in old buckets outside cowsheds. They also make an excellent windowsill plant for winter. Summer dormant, Vallotas start into growth in the autumn and flowering takes place in September before the foliage is fully developed. After flowering, continue watering and feeding until the foliage dies back. Then slowly dry out, but even when fully dormant do not allow to become desiccated. The most splendid of all scarlet lilies. Ht 30cm. Sp. 25cm.

DAHLIAS. (Asteraceae). Dahlias named after Andreas Dahl, a pupil of Linnaeus, are the most gorgeous members of the Daisy family, but the original early 19th Century interest in them was mainly in the tubers as a potential foodstuff. Can you hear it - one portion of Dahlias and chips and make it snappy. Although originating in S. America and Mexico, Dahlias are of easy cultivation and much more hardy than generally supposed. If you are favoured with light soil, like ourselves, they will overwinter in the ground without any protection. For those doing penance for horticultural sins in a previous life, and enduring the rigours of clay, I recommend keeping your Dahlias in pots and plunging them in the herbaceous border. In November you can simply lift the pots and store overwinter under the greenhouse staging. They should be repotted in March with lavish additions of slow release fertilizer; like so many tuberous plants, they are gross feeders. Dahlias, like Chrysanthemums, have attracted much baseless slander. Although there are plenty of vulgar and nasty cultivars, (from which the only gaiety to be extracted is in their demise. "Hurrah.... It is a frost! The dahlias are all dead." Surtees. Handley Cross. 1843). Dahlias, like Arabian Nights, stand up with the proudest of herbaceous plants. They provide an essential barbaric grandeur to the summer and autumn border, (dainty Penstemons are not enough!), flowering profusely from July to November. Plant them with Helenium Moerheim Beauty, Crocosmias, Miscanthus Strictus. Christopher Lloyd recommends them for "peppering up a haze of michaelmas daisies." For admirers of Cosmos atrosanguineus, **D. *Arabian Nights*** is a must. Its fully doubled flowers are of the richest, blackest red. Ht. 120cm. Sp. 40cm. **D. *Beliner Kleene*.** Decorative salmon rose flowers Ht. 50cm. Similar to the illustration from "The New Botanic Garden". **D. *Bishop of Llandaff*** has pillar box, red flowers and almost black foliage. The very lovely illustration by Christine Stephenson does not flatter, the flowers are that sumptuous and the leaves are that dark. Ht. 85cm. Sp.40cm. **D. *Dark Desire*** is my current crush. Raised by Chris Ireland Smith, this is a plant of overwhelming beauty and grace. A chance seedling from the Bishop of Llandaff, it has rejected the grandiloquence of its parent and reverted to the wildling looks of its species

Catanache caerulea *(3)*

ancestors. It has the deepest, darkest, velvety red, single, narrow petalled flowers with the purest gold stamens. The foliage is finely cut and a darkish green. Ht. 90cm. Sp. 60cm. **D. *David Howard*** has deep bronze foliage, which sets off resplendent, orangey yellow flowers. Christopher Lloyd says this dahlia deserves "every scrap of the praise lavished on it." Ht. 75cm. Sp. 40cm. **D. *Engelhardt's Matador*.** Decorative, large cool purple flowers with black foliage. Bob Brown has planted it with Artemisia Powys castle and expresses himself deeply gruntled with the effect. Ht. 100cm. **D. *Grenadier*** – another of Bob Brown's favourites. Very old variety with dark blackish leaves and double scarlet flowers. June-Oct. Ht. 100cm. Sp. 50cm. **D. *Mermaid of Zennor*** has, in the words of Bob Brown, lavender lilac flowers in the shape of a hand with the fingers spread. The petals are long narrow and slightly twisted. Ht. 75cm. **D. *Melody Swing*.** Decorative, orange double flowers. Ht. 50cm. **D. *Moonfire*** has dark bronze foliage and single yellow flowers with a vermillion central ring. Ht. 50cm. Sp. 40cm. **D. *Murdoch*** is an old variety with stunning, exceptionally clear red, double flowers which open from gold bracts. July-Oct. Dark green leaves. Ht. 85cm. Sp. 40cm. **D. *Nuit D'Ete*** has flowers of a similar colour to Arabian Nights, but the flowers are larger, of a small Waterlily type. Ht. 100cm. Sp. 40cm. **D. *Bednall Beauty*** is a compact plant and has deep crimson flowers and elegant black, more deeply divided foliage than the Bishop. Ht. 50cm. Sp. 40cm. **D. *Tally Ho*** has single, vermilion flowers and black foliage. Ht. 75cm. Sp. 40cm. **D. *coccinea*** is a fine species with variable orange, red or yellow single flowers. In Mexico in the wild it can grow to 270cm; perhaps thankfully in English gardens it rarely reaches more than 120cm. with a spread of about 50cm.

DATURA. (Solanaceae). Daturas, or Brugmansias as we are now supposed to call them, are classic terrace plants. Small trees, they look superb in Versailles tubs. Their huge, trumpet shaped bells emit a voluptuous scent in the evening. Although exotic in looks, they are easy to grow. Basic rules - keep them fairly dry in a frost-free greenhouse or conservatory during the winter. During the growing season water daily. Always lace their daily drink with tomato fertiliser; Daturas are incorrigible gross feeders. They should be placed outside as soon as there is no danger of frost. Choose a sheltered position, as their leaves can be torn by wind. Spray the foliage daily, particularly the undersides; this inhibits red spider mite. Daturas should be potted in a John Innes style compost; the weight will prevent them blowing over. Daturas can be cut back brutally in autumn to ease overwintering. **D. *Single Pale Pink*, D. *Single Pale Yellow*, D. *Grand Marnier*** single apricot flowers, **D. *Single White***, and **D. *Knightii*** double white flowers. All of them will grow to about 200cm. tall and 100cm. across.

DELPHINIUM. (Ranunculceae). 'Delphis' being Greek for dolphin and dolphin describing the shape of the flowers. "Yes madam, how about a nice bunch of dolphinflowers?" Liking my Delphiniums good and tall, I prefer the Pacific Giants. I find the modern dwarf hybrids about as pleasing as the miniature Apple tree, Ballerina. The tall, soaring spires of the true Delphiniums are as essential an ingredient to the herbaceous border as Foxgloves and Verbascums. Borders without height are borders without mystery or sky. **D. *elatum*** is a native of Siberia and was introduced early in the 17th Century. The following Pacific Giant Hybrids all descended from Delphinium elatum: **D. *Astolat*** dusky pink, **D. *Black Knight***, perhaps the most popular of all the Pacific hybrids with its darkest blue flowers, **D. *Blue Jay*** clear blue with a dark bee, **D. *King Arthur*** violet blue with a white bee, **D. *Galahad*** white, **D. *Percival*** pure white flowers with a dark eye, and **D. *Summer Skies*** powder blue. They all grow about

Centaurea montana *(1)*

180cm. tall with a spread of 25cm. For those who find the Pacific Giant Hybrids too monumental there are two varieties of **D. *belladonna*. D. *b. Bellamosum*** has dark gentian blue flowers and **D. *b. Casa Blanca*** has pure white flowers. Both grow about 125cm. tall and have a spread of about 40cm. They flower in June-July. Unlike the Pacific Hybrids they do not repeat flower. Delphiniums do not thrive in poor sandy gardens and need a moisture-retentive, fertile soil. They require good drainage and dislike heavy clay in which they are liable to rot. Delphiniums need dividing every three years, as in time their centres begin to decay and the rot can spread to the offsets. Dig them up in early spring, and break off the offsets, which should be replanted individually. The centre of the plant, even if still showing signs of life, should be chucked because its vitality will be spent. **D. *grandiflorum Butterfly*** is a Chinese Alpine which needs sharper drainage. Its intense gentian blue flowers scintillate. Shortlived, but aren't we all, it makes plentiful seed, which is easy to germinate under gentle heat in the spring. All delphiniums need protection from slugs.

DIANTHUS. (Caryophyllaceae). Dianthus comes from the Greek, meaning Zeus' flower. Pinks were probably first introduced into this country from Eastern Europe in the 14th Century by Cistercian monks. Pinks are classic plants for sunny, well-drained, limy soil. They all flower in June-July. **Alpine Dianthus.** All these plants make tight cushions of glaucous foliage - about 10cm. high when in flower and with a spread of about 25cm. Ideal plants for the trough or gravel garden. **D. *Calypso Star*** has raspbery red flowers dappled and edged with pink. Similar to, but more prolific in flower than **D. *Sops and Wine*. D. *Starry Eyes*** has deeply fringed white petals with a rich purple centre. **D. *La Bourbrille*.** Clear pink flowers. Very neat and compact. **D. *Pike's Pink*** has single, very scented, light pink flowers with frilled petals. **D. *Whatfield Wisp*** has single, small, fringed, pale pink, fragrant flowers. **Border Dianthus. D. *barbatus*.** Sweet Williams are native to the South of Europe and are thought to have been introduced into cultivation by Carthusian monks in the 12th Century. Unlike most flowers introduced in the mediaeval period, the Sweet William was prized solely for the beauty of its flowers not for any therapeutic or culinary qualities. "These plants are not used either in meate or medicine, but esteemed for their beautie to decke up gardens, the bosomes of the beautifull, garlands and crownes for pleasure." (Anon). We know that Henry VIII planted them en masse in his new garden at Hampton Court. For crownes, for pleasure or to deck up the bosoms of his many wives? Sweet Williams are short lived perennials and are normally raised from seed. They flower May-June. Ht. 45cm. Sp. 20cm. They require sun and do best in well drained fertile soil. **D. *Gran's Favourite***, a modern laced pink. White flowers laced with purple. The laced pink D. Lady Acland, alas, has disappeared from cultivation. **D. *Mrs Sinkins*** was as Alice Coats tells us "raised in the workhouse; being bred by the Master of Slough Poor Law Institution and named in honour of his wife." It has large, double, white flowers, which are spectacularly scented. Wonderful in a bunch on a bedside table. Ht. 30cm. Sp. 40cm.

DIASCIA. (Scrophulariaceae). Useful for edging. Flower prolifically. June-Oct. Benefits from a hard cutback mid season. Sun and good drainage. Both the following forms are stone hardy and make mats 10cm. high and perhaps 30cm. in diameter. **D. *Blackthorne Apricot*** has warm apricot flowers. **D. *Ruby Field*** has deep pink flowers. **D. *Icecream*,** makes a more substantial plant and has large pure white flowers with dainty pink speckling on the bottom lip. A cracker! June-Oct. Ht. 25cm. Sp. 35cm. All diascias are good in gravel or for providing spot colour at the front of the border. I am

Centranthus ruber *(3)*

no great lover of diascias and agree with Christopher Lloyd when he says of diascias "you don't want too many of this spineless kind of plant or your garden will be in danger of melting away altogether." If you want an alternative to Diascia, try Nemesia d. Confetti. A first rate plant with plenty of structure, closely related to Diascias, it is hardy, and flowers from June till Christmas.

DICTAMNUS. (Rutaceae). English name Dittany. Elizabethan introduction from Europe. Dictamnus got its common name, the Burning Bush, because it is possible on a sunny, still day, when the seed pods are ripening, to ignite the escaping volatile oil by placing a lighted match underneath the seed pods. The oil will burn. The plant will remain intact. Dictamnus is a plant of great elegance, but slow to establish and requiring rich, moist soil. It is important to ensure, that during the early years, it is not overgrown by more vigorous plants. Once established, it is extremely long lived. The foliage has an aromatic lemon scent. The seeds, according to Gerard (I am always forgetting to check the veracity of this) smell of goat. **D. *albus*** has the chastest of pure white flowers. **D. *a. var. purpureus*** has subdued mauve flowers veined with purple. Ht. 75cm. Sp. 25cm.

DIERAMA. (Iridaceae). Dierama, the Greek word for funnel, describes the flower shape; the English name, Angel's Fishing Rod, describes the plant's beautiful habit. **D. *pulcherrium*** throws up 100cm. or more arching stems, which terminate in a string of small pink bell shaped flowers. June-Aug. Smallish rosettes about 20cm. in diameter. **D. *p. Guinevere*** has pure white flowers. **D. *p. Merlin*** has dark, blueberry-purple flowers. **D. *p. Miranda*** is a vigorous clone with soft lilac flowers. They look their best planted by water or gravel, against which the overhanging wands may be admired without distraction; their reflections in the water are also of great beauty. Dierama are extremely deep rooted and therefore need to be planted with care as they resent being dug up and moved. They like a moist fertile, but well drained soil. **D. *igneum,*** a dwarf form which has brick red flowers. July-Aug Ht. 50cm. Sp. 20cm. and **D. *reynoldsii*,** which has wine red flowers and intense silver bracts. July-Aug. Ht. 100cm.

DIGITALIS. (Plantaginaceae). Foxgloves were considered by the early herbalists to be a cure for Scrophula. From this derives the name of the natural order to which foxgloves belong. Foxgloves gained their Botanic generic name from 'digitabulum', the Latin word for thimble. Our common name is, supposedly, a corruption of Folk's glove or Fairy's glove. **D *ferruginea*,** yellow flowers netted with brown. The overall impression is rust as the name ferruginea implies. Tends to be short-lived May-July. Ht. 150cm. Sp. 30cm. **D. *grandiflora* syn. Digitalis ambigua** is perhaps the most reliably perennial of all the foxgloves. It bears beautiful, creamy yellow flowers, heavily blotched inside the bells with maroon, June-July. A native to the European Alps, it was already being cultivated by Gerard in his London garden in the 16th Century. It prefers a more open situation than Digitalis Purpurea; its height is 60cm. Its spread about 25cm. **D. *mertonensis*** is a hybrid between Digitalis purpurea and Digitalis grandiflora. It bears flowers the colour of crushed strawberry in June-Aug and grows to about 60cm. in height, Sp. 35cm. It will take light shade or full sun and makes splendid rosettes of leathery, dark green leaves. Unlike Digitalis grandiflora, which should be left undisturbed for years, Digitalis mertonensis benefits from division every couple of years, immediately after flowering. The biennial forms of Digitalis purpurea are described in the section for shade loving plants.

ECHEVERIA. (Crassulaceae). Echeverias are named after a Mexican 19th Century botanist, Atnasio Echeverria. **E. *Black Prince*** has very dark purple almost black rosettes

Ceratostigma plumbaginoides *(1)*

with salmon red flowers in the fall. Rosettes about 8cm. in diameter. E. ***elegans*** is a lovely plant with large blue, fleshy rosettes and pinkish orange flowers. It looks wonderful established in a large clay pan, its rosettes spilling over the edge. Ht. 40cm. Sp. 50cm. E. ***Duchess Of Nuremberg*** has grey rosettes flushed with pink. It is similar in habit to E. elegans and has cream flowers.

ECHINACEA. (Asteraceae). 'Echinos' is Greek for hedgehog – a reference to this plant's prickly phyllaries. "Beware of the Prickly Phyllaries" sounds like a line out of Edward Lear. Echinaceas, like so many North American natives are robust hardy plants thriving in heavier soils. They share with Rudbeckias, to which they are closely related, the English common name Coneflowers. All flower from July to September. E. ***Arts Pride*** is an exciting form with orange red flowers, a hybrid between E. purpurea and E. paradoxa. Ht. 60cm. Sp. 30cm. E. ***paradoxa Mellow Yellow*** has soft yellow flowers above dark green foliage. Ht. 70cm. E. ***purpurea Magnus***, has deep cerise flowers with orange brown central bosses - looks very good as a buttonhole with a loud check suit. E. ***p. White Swan***, has white flowers which are too tasteful for my sartorial requirements. E. Magnus is, as his name suggests, a hefty fellow, growing to over 90cm. E. White Swan is a daintier 75cm. Both have a spread of about 25cm. E. ***p. Kim's Knee High***, has clear bright pink flowers. Useful for its lack of stature, - as is its white flowered sport E. ***p. Kim's Mophead***. Both have a height of 55cm. and a spread of 25cm. Perhaps my favourite Echinacea is E ***p. Ruby Giant***, which has large ruby red flowers with a prominent bronze boss. The most admired new plant on the nursery in 2004. Ht. 85cm. Sp. 25cm. E. ***teneseensis Rocky Top Hybrids***. According to Bob Brown a strong new hybrid from a previously miffy species. Cerise-pink flowers. July-Oct. Ht. 60cm. Sp. 25cm. All echinaceas must be guarded against slugs. They look good planted with grasses and their seed heads make a persistent feature in the winter.

ECHINOPS. (Asteraceae). E. ***ritro Veitch's Blue*** has deep blue flowers June-August and grows to 80cm. Sp. 40cm. E. ***sphaerocephalus Arctic Glow*** has larger white flowerheads. Ht. 80cm. Sp. 40cm. Native to the Mediterranean, Echinops thrive in poor dry soil. Fertile moist soil is best avoided, as plants will need staking. Bowles writes: "I have often recommended them for entomologists' gardens, where plants are wished for that can be visited after dark with a lantern to surprise a supper party of noctuid (sic) moths."

ECHIUM. (Boraginaceae). E. ***russicum***. Dark chocolate red flowers in June. Silvery grey leaves Ht. 40cm. Sp. 30cm. Sun. Well drained sandy soil. Hardy to -10 C.

EPILOBIUM. (Onagraceae). E. ***angustifolium*** The Rosebay Willowherb, is a British native but until the 20th Century was considered a choice rarity - somewhat puzzling because the plant has both a vigorous stoloniferous root and is a prodigal self seeder. White in his 'Natural History of Selborne' records, on one botanical sortie, gathering armfuls of wild orchids and to his great delight a few stems of willowherb. John Hill writing in 1756 tells us "The garden does not afford a plant more specious and elegant than this, although it be a native of our country. It is not so common wild as to put the vulgar in mind of calling it a weed in gardens; and those who have judgement despise the little prejudices which represent everything as mean that comes easily". Admirable unsnobbish sentiments, but it is a plant to be feared as Tournefort writing in 1709 says, for its "licentious and ungovernable creeping". Charles Bryant writing in 1783 advises us that the young tender shoots cut in spring are little inferior to Asparagus. Willowherb leaves were apparently sometimes added to Russian teas to give them a China flavour.

Cerinthe major *(3)*

E. ***angustifolium Album***, the choice white form is not nearly as invasive as the species. White flowers. June-Sept. A most beautiful elegant plant. Ht. 110cm. Sp. 70cm. Sun or part shade. Not too dry soil.

EREMURUS (Liliaceae). Eremurus are Asian cousins of the Mediterranean Asphodelus. They have the same fleshy roots and albeit it on a grander scale and with far more delicate colouring the same close packed spires of flowers. Eremurus require full sun and thrive in chalky soils. They flower in June-July by which time their rosettes are already withering. Plant Eremurus among bearded irises (they enjoy the same conditions) and they will provide colour in summer, when the irises have finished flowering. E. ***x isabellinus Image*** has white flowers with protruding gold stamens. Ht. 120cm. Sp. 30cm. E. ***robustus*** grows 250cm. tall and has pink flowers. E. ***stenophyllus*** has dark yellow flowers with orange stripes and grows 175cm. tall. They all flower in June-July.

ERIGERON. (Asteraceae). The name Erigeron derives from the Greek for an old man. This, so I am told refers to its hairy pappus. "You great hairy pappus" has a fine obscene ring to it, but unfortunately pappus translates mundanely into commonal garden language as the ring of fine hair, which grows above the seed and expedites seed dissemination. The two following Erigerons, which are of N. American origin, thrive in any reasonably fertile soils. E. ***Dunkelste Aller***, as its name suggests has the darkest of all violet flowers. E. ***Snow White*** has pure white flowers. Both have a height of about 60cm. and will make a spread of about 40cm. Both flower June-August. For those with gardens of impoverished sand, E. ***glaucus***, native to Turkestan, is a valuable, totally drought resistant, mat forming plant. It is completely resistant to salt spray and makes a wonderful plant for covering a dry sunny bank. E. ***g. Sea Breeze*** has bright pink flowers and looks splendid in marine gardens with the perennial white stock. Will grow and thrive in pure sand. June-Sept. Ht. 20cm. Sp. 60cm. E. ***karvinskianus*** is a petite member of the genus. Masses of small delicate white daisies, which age to pale pink. An airy beauty, which self-seeds into walls and paving. Ht. 20cm. Sp. 25cm.

ERODIUM. (Geraniaceae). The English name for this plant is Storksbill. It is related closely to the Geranium or Cranesbill family. The bills refer to the shape of their seeds rather than their extravagant habits. In fact the reverse is true. Like Cranesbills, Erodiums are undemanding plants and thrive in poor sandy soil. They look at ease in the gravel garden. E. ***chrysanthum*** is a dainty plant in all respects. Filigree silver foliage. Exquisite, pale yellow flowers. June-July. Ht. 25cm. Sp. 25cm. E. ***manescavi***, a larger plant, has umbels of wine red flowers June-Sept, a plant of perfect poise – Ht. 30cm. Sp. 25cm. E. ***petraeum*** is a plant of great charm. It has finely divided ferny foliage and dainty white flowers with maroon splodges. May-August. Ht.. 15cm. Spread 25cm. E. ***reichardii*** is a delightful miniature. Ht. 5cm. Sp. 25cm. It makes a compact mat studded with pinkish red flowers June-Sept. E. ***reichardii Alba***, the lovely white form is identical in habit and flowering season. E. ***trifolium***, (previously erroneously called E. pelargoniflorum), is a plant for all seasons - in winter it makes a handsome mound of fresh green foliage, which is surprisingly resistant to frost, and in spring and summer gives a ceaseless display of dainty white flowers, striated with pink. Ht. 30cm. Sp. 30cm. Both E. manescavi and E. trifolium discreetly self-seed.

ERYNGIUMS. (Apiaceae). In the 17th Century E. maritimum was considered by herbalists to be an elixir of youth, and the town of Colchester had a thriving industry, manufacturing Eryngium tablets for the sick and the elderly. Sea holly was also in

Chelone obliqua *(3)*

demand as an aphrodisiac; Jack Falstaff, for whom drink increased the desire, but took away the performance, makes mention of it. Sea hollies are the aristocrats of the thistle world. In shape their flowers resemble Echinops, but there any likeness stops. Eryngiums are entirely non invasive. Would that the same could be said of Echinops! Eryngiums' foliage is dramatic and handsome. The best that can be said about the foliage of Echinops is that it is drab! Eryngiums are easy and accommodating plants. They need a deep and well drained soil, and thrive in full sun. Being tap rooted, they resent disturbance and propagation is best accomplished by seed. Eryngiums readily appeal to those who like all things horrid and prickly, being plentifully supplied with spikes and barbs. One of the nastiest of the lot is called **E. *horridum***. This has evergreen, glossy barbed foliage and sends up an astonishing 240cm. stem, terminating in a candelabrum of whitish green flower heads in July-August. Sp. 50cm. Long lasting in flower. At the other end of the scale is **E. *variifolium.*** Of all the Eryngiums, E. variifolium is perhaps the most demure. Its basal rosette is of a rich glossy dark green, handsomely veined with white. It is evergreen and well worth growing for its winter interest alone. Silver blue flowers in July-August. Ht. 40cm. Sp. 20cm. It looks very showy planted with Ophiopogon planiscapus Nigrescens, the black lily grass. **E. *alpinum*** has the largest flowers of all the species eryngiums. Blue flowers. Blue stems. May-August. Ht. 75cm. Sp. 25cm. **E. *bourgatii*** flowers June-Aug. Ht. 55cm. Sp. 25cm. **E. *b. Graham Stuart Thomas*** has much the daintiest foliage of all the E. bourgatii. Intense blue flowers. Highly regarded by that great plantsman Bob Brown. **E. *b. Picos Amethyst*** is a form with electric violet flowers. **E. *b. Picos Blue*** has intense blue flowers. **E. *planum***, a native of Southern Europe was in cultivation in England by 1648. **E. *p. Blaukappe*** has very much more intense blue flowers and stems than the species, which can be somewhat insipid. July-September. Ht. 120cm. Sp. 25cm. ***E. maritimum***, has small dark blue flowers, surrounded by large silver bracts. July-Aug. Its most striking feature is its prickly silver foliage which is wax coated to reduce transpiration. Ht. 45cm. Sp. 50cm. E. maritimum is totally drought and salt tolerant. It is native to the British coastline and grows in the shingle banks at Walberswick. Geoffrey Grigson says that "its roots grope through the sand for five or six feet." **E. *tripartitum*** has widespreading heads of blue flowers. July-Aug. Nicely cut basal foliage. G. S. Thomas recommends planting it behind Crambe maritima. Ht. 70cm. Sp. 60cm. **E. *x oliverianum*** has large heads of blue flowers and bright metallic blue stems. Ht. 60cm. The finest display of Eryngiums I ever saw was in a garden composed entirely of sea pebbles about the size of large ducks' eggs. Very nasty indeed it looked. Smooth, hard, unremitting stones provided just the right callous background for the Eryngiums' thorny beauty. The scene only lacked a few towering Onopordums to give it the feel of some infernal, horticultural armoury. Other plants with which Eryngiums consort well are Yuccas and Cynara cardunculus, the Cardoon. All Eryngiums are beloved by practitioners of that arcane species of necrophilia, dried flower arranging.

ERYSIMUM. (Brassicaceae). Erysimums are members of the Cabbage family, although with their narrow lanceolate foliage and shrubby habit they bear no resemblance to our culinary Brassicas. The name Erysimum derives from the Greek word for to draw out, some species being capable of producing blisters. The English name Perennial Wallflower is apt because the Erysimum are often seen growing out of old walls, living apparently on thin air and lime mortar. In the garden they need poor dry soil; in rich soil they seldom survive more than a season. All Erysimums are short lived perennials,

Cimicifuga racemosa *(3)*

losing their vigour as they become woody with age. They can be easily renewed from softwood cuttings. E. ***Apricot Twist***, apricot and mauve flowers. April-Sept. Grey green foliage. Ht. 60cm. Sp 30cm. E. ***Bowles Mauve*** in a sheltered position produces a never ending succession of purple flowers throughout the year. Grey green leaves. Makes a very handsome, perfectly spherical plant. Ht. 80cm. Sp. 80cm. E. ***Chelsea Jacket***, yellow flowers which mutate to pink. May-July. Ht. 60cm. Sp. 50cm. E. ***Cream Delight*** was my plant of the year for 2003. Its soft creamy yellow flowers are the largest, most scented and most beautiful of any Erysimum I know. Green foliage. Ht. 35cm. Sp. 30cm. E. ***cheiri Harpur Crewe*** is a Tudor survival. It produces a profusion of the most enchanting scented, double, deep yellow flowers April-June. Ht. 50cm. Sp. 30cm. Perhaps Francis Bacon was thinking of Harpur Crewe in his essay "On Gardens", when he wrote that wallflowers because of their scent "were very delightful to be set under a parlour or lower chamber window." E. ***Cotswold Gem*** is an excellent, vigorous, variegated Erysimum. Much better than the old E. linifolia Variegata. Tan and pale purple flowers over yellow variegated foliage. April-Oct. Ht. 50cm. Sp. 50cm.
EUCOMIS. (Liliaceae). Eucomis are called the Pineapple Lily after the tuft of leaves which surmounts a closely packed barrel of flowers. E. ***Sparkling Burgundy*** has the darkest burgundy leaves and deep red flowers on deep red stems. Aug-Nov. Unusually the leaves lose none of their colour during the summer and autumn. A dramatic plant, which looks good with the late flowering Aconitum Arendsii. Eucomis although native to S. Africa are stone hardy. They need full sun and will grow in any well drained soil. Their foliage emerges very late in the season so the dormant bulbs should be carefully marked.
EUPHORBIA. (Euphorbiaceae). Named after Euphorbius, Greek physician to King Jubus II of Mauritania, who in the 1st Century A.D. used the milky sap or latex for medicinal purposes; one hopes not as an eye ointment. For some people the latex can cause extreme irritation, and therefore gloves should be worn when handling this plant. The Euphorbia family is truly vast, comprising over 2,000 species, most of which are not suitable for cultivation in this country, being succulents requiring desert conditions. The Euphorbias, listed below, are all Mediterranean in origin and have more normal needs, thriving in any well-drained soil. E. ***ceratocarpa*** originates in Sicily and S. Italy. Large heads of golden flowers. June-Oct. Ht. 150cm. Sp 150cm. No other Euphorbia holds the colour of its flowers over such a long period. It makes a wonderful loose cloud of billowing gold, the very opposite of such highly structured plants as E. ***Wulfenii***. Needs sunny sheltered position. E. ***characias*** has beige bracts with black anthers, much more subtle than the gaudy yellow flowers of the subspecies E. c. wulfenii, which, by comparison, seem blowsy and in your face. Also I think, the narrow glaucous leaves of E. characias superior to the broader, cruder, green leaves of E. c. wulfenii. E. characias flowers Feb-April and grows to about 90cm. in diameter making large architectural plants, which look their best sprawling onto gravel or paving. There is an old churchyard in Norwich entirely furnished with E. characias and Helleborus argutifolius; the strong architectural shapes of the plants against the weathered stone, the quiet greens and creams are a textbook in the value of gardening with a limited palette. E. characias grows quite happily in shade (though not under trees) as well as sun. It benefits from having the old flowering canes cut down in midsummer, by when next year's flowering stems will be beginning to shoot from the base. E. ***c. Black Pearl*** is a short form, which has very prominent black anthers set off by bright green bracts.

Cistus ladaniferus *(3)*

Ht. 60cm. Sp. 40cm. E *cyparissias Fen's Ruby* has startling red shoots in spring. Dusky purple foliage. Honey scented yellow flowers in June. Spreads stoloniferously. Ht. 25cm. Sp. 60cm. The leaden green shoots of E. ***griffithii Fireglow*** emerge in spring washed with orange which slowly fades as the leaves unfurl. In May-June, it produces flowers with the most astonishing orange bracts. Needs soil with some moisture. Sun or part shade. Ht. 75cm. Sp. 60cm. **E. *mellifera*** is the largest Euphorbia we can grow in this country. Native to the Canaries and Madeira, it resembles in shape a giant Euphorbia wulfenii; its foliage is, however, a much fresher green in colour. In the wild Euphorbia mellifera grows up to 1500cm. in height. In England it never reaches such tree like proportions, at most reaching 450cm. and indeed it is normally kept to a modest 150cm. by cutting down the old stems after flowering. Extremely rare now in the wild, Euphorbia mellifera is not difficult to grow, provided it is given a warm sheltered position. In colder parts of the country this means a south or west wall. In milder areas it will grow in open ground. If cut to the ground in severe winters, it will normally reshoot from the base in spring. It likes rich fertile soil. The scented flowers are much loved by bees. Euphorbia mellifera like all Euphorbias can grow rangy with age; as the stems grow up, they tend to lose their lower leaves, looking increasingly gaunt. Ruthlessly pruned in late spring after flowering, plants will renew themselves from basal buds. E. ***myrsinites*** in March-April bears greeny yellow flowers, which fade to pink. It is for its magnificent foliage that this plant is grown. A well established plant makes a glorious glaucous heap. Stems up to 50cm. long, crowded with pale blue leaves along their full length, spill out in profusion from a central crown. Ht. 20cm. Sp. 100cm. Splendidly metallic in gravel with Eryngiums and Onopordums. **E. *palustris*.** Graham Stuart Thomas calls this "one of the most spectacular of spring flowering plants." Brilliant yellow flowers open in April and persist throughout the summer. Autumn leaves, which are a warm yellow, fall to reveal brilliant pink stems in October. Despite its name, (palustris is Latin for marsh), E. palustris also grows happily in dry soils. Ht. 90cm. Sp. 30m. **E. *rigida*** is like a more upright form of E. mellifera. Pointed blue leaves. Greenish yellow flowers in April-May. Ht. 40cm. Needs a warm sheltered spot. **E. x *martinii*** bears lime green flowers in May-June. It has evergreen, very dark reddish black foliage and red stems. Ht. 60cm. Sp. 30cm. Compact. Long lived and well behaved. **E *polychroma Lacy*.** Green leaves with wide cream margins. Pale creamy yellow flowers March-May, which look good in their withering. Ht. 50cm. Sp. 30cm. **E. *Whistleberry Garnet*.** E. amygdaloides robbiae x E. martini hybrid. Very compact. Evergreen. Young foliage shoots rich purple. Sulphur yellow flowers April-May. Ht. 60cm. Sp. 40cm.

FOENICULUM. (Umbelliferae). **F. *vulgare purpureum*** a delightful misty giant of a plant. Covered with the finest, deep bronze, hair-like leaves, it makes large branching candelabra some 180cm. tall, topped with yellow flowers April-September. Sun or part shade. Drought tolerant.

FRAGARIA. (Rosaceae). **F. *Lipstick*.** A hybrid from the garden strawberry and Potentilla palustris. Vivid rose flowers April-May and then again in September. Rich deep green foliage. Good ground cover for moist fertile soil in sun or part shade. Also good in hanging baskets.

FRANCOA. (Melianthaceae). **F. *ramosa*.** Maiden's wreath. Lyre shaped leaves. **F. *Confetti*.** bears white flowers on long dark pink stems. July-Sept. Very chaste and elegant. Ht. 75cm. Sp. 30cm. Any well drained not bone dry soil. Francoa, because of its

Convallaria majalis *(3)*

dramatic foliage, makes a superb specimen in a large pot and is used thus at Tintinhull by the pool.
FRITILLARIA. (Liliaceae). **F. *imperialis.*** "Lily of the Turbaned countries." The crown imperial was introduced into England in the latter part of the 16th Century and has been a frequent subject of botanical illustration ever since, so magnificent is it with its majestic hanging bells and green topknot. Our two illustrations of F. imperialis demonstrate how even botanical illustrations are cultural constructs. The illustration from the New Botanic Garden (published 1812) portrays the Crown Imperial in gorgeous majesty – it could almost be a design for a chandelier at The Brighton Pavilion. The illustration by John Nash (published 1948) emphasizes the wildling beauty of the plant. Both visions are partial truths in our search for the identity of The Crown Imperial! Crown Imperials smell strongly in all their parts. Shove a fork in the ground and pierce one of their bulbs, which incidentally will do them no harm, and you will be enveloped in a wonderful, intoxicating sexy, aroma. Not all agree. E.A. Bowles writes: "they possess such an awful stink, a mixture of mangy fox, dirty dog kennel, the Small Cats' house at the Zoo, and Exeter Railway Station, where for some unknown reason the trains let out their superfluous gas to poison the travellers." **F. *i. Rubra,*** red, and **F. *i. Lutea*** the rarer, beautiful yellow. Both are happy in sun or light shade; they thrive in any soil, except heavy clay or bog, needing a fairly dry rest period in the summer. May-June. Ht. 60cm. Sp. 25cm. If your soil is likely to stay wet, you should dig the bulbs up when the flowering stems have withered. The bulbs can be stored and replanted in late autumn. Crown Imperials look their best planted against a dark background, such as a yew hedge, or an old tarred shed. A couple of the Crown Imperial's dwarf cousins: **F. *pontica*** has grey green bells with brown markings. April. Ht. 30cm. Sp. 15cm. Sun. Loamy soil, enriched with leafmould, which does not dry out in summer. **F. *uva vulpis.*** Bells purple brown on the ouside with golden brown interior. The recurved lips of the bells are yellow. April. Ht. 35cm. Sp. 10cm. Moist rich soil. Good for naturalizing.
FUCHSIA. (Onagraceae). **F. *loxensis.*** Elegant long narrow orange bells. July-Oct. Large mid green leaves. Extremely exotic looking but stone hardy. Ht. 40cm. **F. *magellanica,*** the species plant has purple and red flowers and **F. *magellanica Hawkshead*** has pure white flowers and deep green foliage. Both flower July-Nov and are hardy. Ht. 100cm. Sp. 80cm.
GAILLARDIA. (Asteraceae). The Jam Tart plant. Most Gaillardias are annuals. **G. *grandiflora*** is a true perennial. **G. *g. Burgunder*** has sumptuous dark red flowers. June-Oct. Ht. 50cm. Sp. 30cm. Gaillardias are victims of misplaced snobbery, as for that matter are jam tarts. They are such willing cottage garden plants, which flower unstintingly through the long summer months. Like Delphiniums, Gaillardias need regular division. They should be dug up in early spring, their offsets broken off and replanted, and the exhausted central portion discarded. Any not too dry soil.
GALANTHUS. (Amaryllidaceae). The Latin name coined by Linnaeus means milkflower. Snowdrops are not native to Britain. They were probably first introduced from Europe by monks during the mediaeval period. G. nivalis is native to Southern and Central Europe. Their flowers were traditionally used for decking the Lady Altar on the feast of the Purification of the Virgin. Galanthus only gained their English name "Snowdrop" in the 17th Century. Previously they were referred to as "bulbous Violets". Snowdrops close their flowers at night to encapsulate the daytime warmth; the temperature inside the bell is sometimes as much as two degrees warmer than the surrounding air. **G. *elwesii.*** Large snowdrop with broad glaucous, slightly twisted

Cyclamen persicum *(3)*

leaves. Outer segments of flowers pure white. Inner segments have striking green markings. February. Native to Turkey. Best in cool shade. Ht. variable 10 - 25cm. Sp. 10cm. G. ***nivalis Viridaprice.*** White flowers with green markings on both inner and outer segments. February. Ht. 16cm. Sp. 10cm. Sun or part shade. Any not too dry soil. G. ***woronowii.*** Native to Georgia Russia and Turkey. White flowers. Inner segments with green markings. Late flowering. February-March. Glossy deep green leaves. Ht 10cm. Sp. 10cm. Cool moist soil. Full or part shade.

GALEGA. (Leguminosae). G. ***officinalis.*** The name Galega comes from the Greek for milk 'gala'; eating the plant was once thought to increase lactation. Its English name, Goats Rue, is no more accurate, Galega being a member of the pea family. Once a standard cottage garden plant, fashion has now largely banished it from our borders. It is a splendid robust plant, in appearance perhaps more like a shrub than an herbaceous plant. Strong, erect, stems emerge from its crown, handsomely furnished with typical, vetch like foliage. In summer it bears in profusion small, bicoloured mauve and white flowers. June-Sept. Ht. 70cm. Sp. 60cm. Grow in full sun - in part shade, tends to be floppy. Relishes clay soils. Galega is one of those understated flowers which I cherish for its lack of ostentation. Not so Christopher Lloyd who, like a man addicted to strong liquor, must have strong colour. Of Galega he says disparagingly. "I like to see it at its best - in *your* border."

GALIUM. (Rubiaceae). G. ***verum*** is our native Lady's Bedstraw. Honey scented gold flowers. June-September. Ht. 40cm. Sp. 60cm. Like Woodruff, Lady's Bedstraw, when dried, smells of hay and was commonly used for stuffing mattresses. A North European mediaeval legend recounts that Mary during the Nativity lay on a litter of Lady's Bedstraw and Bracken. G. verum needs well drained soil and full sun. It has a running root and will naturalise itself in grass. Robert Turner in his Botanologia 1664 writes "The decoction of the herb and flowers used warm is excellent to bathe the surbated Feet of Footmen and Lackies in hot weather." I guess relief might also be sought by superannuated Nurserymen.

GALTONIA. (Asteraceae). G. ***candicans*** Large heads of cool white flowers in August. G. Stuart Thomas suggests planting among paeonies to provide summer interest. Ht. 120cm. Sp. 40cm. G. ***princeps.*** White bells tinged with green on erect stems. July-Sept. Sun. Good drainage. Ht. 30cm. Sp. 20cm. G. ***regalis.*** Pendulous greenish yellow flowers. July-Sept. Ht. 80cm. Sp. 30cm. G. ***viridflora.*** Greenish white pendulous bells. July-Sept. Ht. 60cm. Sp. 30cm. The green Galtonias look lovely planted with Anthericum ramosum and white Agapanthus. Christopher Lloyd, preferring the overstated showy plant to gently understated elegance, says of G. princeps "I like it but would prefer a good Eucomis."

GAURA. (Onagraceae). G. ***lindheimeri.*** Gaura lindheimeri is native to the South East United States, and although sometimes described as half hardy, this can only be said to apply to gardens of unimproved, heavy clay. In well drained soil, protected from icy wind, Gaura survives most winters. It is however shortlived and plants should be renewed every three years. Sometimes called the butterfly plant, Gaura is a pleasing and most floriferous plant. The whole plant is well furnished with elegant, willow like foliage. The flowers, small and white with pink bracts, are produced, prolifically, throughout the summer and autumn. Because of its abundance of small flowers, Gaura lindheimeri is invaluable for bringing lightness and froth to the late flowering border. I like to plant it with Asters, Dahlias, and Chrysanthemums, which, unalleviated, can begin to look like so many show jumping rosettes. It also looks good

Cynara cardunculus *(2)*

with grasses, its stems moving gently in the breeze. Ht. 100cm. Sp. 40cm. **G. *l. My Melody*** is a good variegated form, with nice bright gold markings on the leaves; it has white flowers and is just as free flowering as the species but has the advantage for front of border planting of being much more compact. Ht. 35cm. Sp. 30cm. **G. *l. Siskiyou Pink*** is a delightful deep pink variety of medium height. Red leaves. Lovely planted with Stipa calamagrostis. Ht. 50cm. Sp. 50cm. **G. *l. Cherry Brandy*** is another pink flowered variety with red flushed foliage. Very compact. Much admired in our garden. Ht. 35cm. Sp. 45cm. Christopher Lloyd is predictably suspicious of Gauras. He issues the warning: "beware of harbouring too many plants in your garden of which the adjectives graceful and charming spring to your besotted lips." He suggests, as an antidote, large lashings of Kniphofias and Agapanthus.

GENTIANA. (Gentianaceae). The gentian was named after Gentius King of Illyria (181-168 BC), who was supposed to have discovered its medicinal properties. It is still used today as an antiseptic (Gentian Violet) and as a tonic. **G. *acaulis*** is native to the European Alps. Deep blue flowers. May-June. Ht. 10cm. Sp. 20cm. **G. *asclepiadea.*** The Willow Gentian. Blue upward facing flowers sprout from the axils of the upper leaves. Aug. Ht. 60cm. Sp. 60cm. Any moist fertile soil. Full or part shade. Reginald Farrer captures perfectly (as always) the habit of the Willow Gentian: "inclining this way and that beneath the burden of its beautiful sapphire trumpets. String and stakes must not be allowed near this wayward beauty." **G. *dahurica.*** Tufted plant with fairly erect stems. Deep blue flowers, which are paler in the tube. June-July. Ht. 20cm. Sp. 25cm. Sun or part shade. Any not too dry soil. **G. *triflora.*** Upstanding plant with blue flowers carried in the upper axils. Aug-Oct. Ht. 90cm. Sp. 50cm. Moist peaty soil. Sun or partial shade. **G. *verna.*** Native to Teesdale. Deep blue flowers. April-May. Ht. 10cm. Sp. 30cm. Limelover. Open position.

GEUMS. (Rosaceae). Geum comes from the Greek Geuo, meaning tasty – a reference to the aromatic roots of G. urbanum. Geums as G. S. Thomas says "are some of the gayest of early summer plants". They are rich in colour and prolific in bloom. They have handsome, deep evergreen foliage, and bear their flowers on strong stems. They are easy in cultivation, given fertile, moisture retentive soil. They are happy in either full sun or part shade. They do well in clay soils. **G. *Beech House Apricot*** has large apricot orange flowers May-July. Originated in the Shackletons' fabled garden in Ireland. Ht. 30cm. Sp. 30cm. **G *chiloense Dolly North*** has orange single flowers. Ht. 40cm. Sp. 30cm. **G. *c. Lady Stratheden*** has wonderful, rich yellow, double butterball flowers. Ht. 60cm. Sp. 40cm. **G. *c. Mrs Bradshaw*** has scarlet double flowers. Looks wonderful with Anchusa Loddon Royalist. Originated as a chance seedling in a bed of G. chiloense. Ht. 60cm. Sp. 40cm. **G. *montanum* Borisii** bears bright tangerine orange flowers from May to September, well shown off by bright green leaves. Ht. 30cm. Sp. 50cm. **G. *rivale Leonards Variety*** has soft old rose bells with dark maroon calyxes from May to July. Christopher Lloyd describes the flower colour as "sad", which coming from the Great Dixter Guru, you can interpret as meaning "not in your face". Ht. 30cm. Sp. 40cm. **G. *x Princess Juliana.*** Clear orange flowers on elegant dark stems. Floriferous. Perhaps the best of the orange geums – all the other orange varieties I know seem to be afflicted with coarse stubby stems. May-Sept. Ht. 45cm. Sp. 75cm. Good with dark leaved Heucheras. **G. *x. Rijnstrom*** has coppery orange flowers May-August. Ht. 50cm. Sp. 30cm. **G *x Werner Ahrends*** is a short form, which bears orange red flowers May-August. Ht. 30cm. Sp. 25cm.

Dahlia Arabian Nights *(2)*

GILLENIA. (Rosaceae). **G. *trifoliata*** is one of the most elegant of all herbaceous plants. Native to the U. S. A. it is a wiry plant, about 100cm. in height and 60 cm. in spread. G. trifoliata was first introduced to England by Bishop Compton, who grew it in his garden at Fulham Palace. Its reddish stems, which carry the minimum of foliage, bear a delicate scattering of white flowers in June. The white flowers are enhanced by red calyxes, which persist long after the flowers have finished. The foliage colours well in autumn. Humus rich, acid or neutral, woodland soil. Good drainage. Part shade.

GLADIOLUS. (Iridaceae). Gladioli get their name from the Latin word 'gladius', meaning sword, a reference to the shape of the leaves. All gladioli need well drained soil and full sun. **G. *byzantinus***. Cultivated by John Parkinson in 1629. Magenta crimson flowers in May-June. Ht. 60cm. Sp. 20cm. Sun. Well drained light soil. Native to the Mediterranean, G. byzantinus is found growing wild in cultivated fields, its corms being deep enough to elude the plough. **G. *cardinalis***. Specatacular crimson flowers with white splashes. July. Bluish leaves. Ht. 60cm. Sp. 25cm. winter grower. Hardy in our garden. Looks wonderful growing through Sedum Purple Emperor. **G. *Flevo Junior***. Richest velvety red flowers July. Ht. 60cm. Sp. 20cm. **G. *murielae* syn. Acidanthera bicolour**. Night scented white flowers with purple spot at the base of the petals. Aug-Sept. Sun. Well drained soil. Ht. 90cm. Sp. 30cm. Tender. Lift in autumn. **G. *nanus albus***. We are told this is a better plant than G. The Bride, most stocks of which are virus ridden and perform indifferently. White flowers. Ht. 45cm. Sp. 20cm. **G. *papilio***. Creamy hooded flowers smudged with purple and green. Aug-Sept. Narrow grey green leaves. Multiplies rapidly. The only gladiolus I know which does well in heavy soil. Ht. 90cm. Sp. 25cm. **G. *Purple Prince***. Black buds open to deepest violet, medium sized flowers. **G. *tristis***. Cultivated in the Chelsea Physic Garden in 1745. Delicious, night fragrant, cream flowers in May-June. Ht. 60cm. Sp. 20cm. Sun. Good drainage and not too much frost. Quite, quite beautiful. As Christopher Lloyd remarks, G. tristis flowers quickly lose their scent when cut, so you should grow some in pots in the greenhouse to bring into the house.

GLAUCIUM. (Papaveraceae). **G. *flavum***. The yellow horned poppy. Glaucous leaves. Gold Chinese silk flowers June-July. Handsome curved seedpods in late summer and autumn. Ht. 50cm. Sp. 35cm. Grows in the sand dunes between Dunwich and Southwold. Resents root disturbance and needs sharp well drained soil. Good in the gravel garden where it will readily self seed.

GYPSOPHILA. (Caryophyllaceae). **G. *cerastiodes***. White flowers veined with purple. May-June. Fresh green hairy leaves. Forms a small cushion. Ht. 8cm. Sp. 20cm. **G. *repens*** forms semi-prostrate mats of grey green foliage, covered with flowers throughout May and June. Good in a trough or the gravel garden. **G. *r. Rosea*** has mid pink flowers and **G. *r. alba*** has chalk white flowers. Sun. Good drainage. Ht. 20cm. Sp. 30cm. **G. *paniculata Bristol Fairy*** produces a wonderful froth of small double white flowers. July-Aug. Ht. 100cm. Sp. 100cm. Gertrude Jekyll suggests planting it behind spring bulbs and Oriental Poppies to hide the void in summer. Needs plenty of sun, space and a well drained soil.

HALIMIOCISTUS. (Cistaceae). **H. *revollii***. Broad rounded leaves smothered in June-July with small, white, cup shaped flowers with gold centres. Makes a small rounded shrub. Ht. 50cm. Sp. 50cm.

HELENIUM. (Asteraceae). **H. *autumnale***. As G. S. Thomas so rightly says, Helenium autumnale need careful planting. They need to be kept away from the pink flowered

Dahlia Bishop of Llandaff *(2)*

plants of summer, the Sidalceas, Phloxes, Monardas and Lythrums. They look their best planted with Dahlias and Crocosmias. They also need to be planted with good foliage plants as their own verges on the disgraceful. Large clumps of Hemerocallis in the foreground to disguise their bony ankles – the late flowering H. Chicago Apache with its red brown flowers in my garden. Helenium autumnale is easy in cultivation; given moist soil. It will grow in either full sun or part shade. Like so many Compositae, it is North American in origin. **H. *a. Moerheim Beauty*** has brownish red flowers, and **H. *a. Dunkelpracht*** has deep red flowers which age to brown. Both flower from July to September. Both grow to about 80cm. with a spread of 40cm. **H. *a. Sahin's Early Variety.*** Streaked orange flowers. Not only does it start earlier but also it goes on longer than other varieties. June-Oct. Ht. 100cm. Sp. 40cm. Beware! Slugs love Heleniums even more than they do Hostas. Heleniums, like most Compositae, have until recently been out of vogue but thanks largely to Piat Oudolf things have changed and they are at last attracting the interest of plant breeders.

HELIANTHEMUM. (Cistaceae). Rockroses need to be chosen with care. Some of the more unusual varieties, such as Raspberry Ripple, are sad, puny creatures, which will only ever malinger. The following varieties have plenty of vigour. They all flower June-July and grow to about 35cm. high and 60cm. or more in spread. **H. *Ben Hope*** has carmine pink flowers. **H. *Henfield Brilliant*** has superb, burnt orange flowers and looks wonderful with Lithodora Heavenly Blue. **H. *Wisley Pink*** is a lovely, soft pale rose. **H. *Wisley Primrose*** bears tasteful, pale yellow flowers. Helianthemums look at home in the gravel or rock garden. They all need cutting back hard after flowering to keep them compact and encourage fresh new growth.

HELIANTHUS. (Compositae). The perennial sunflowers are useful upstanding plants to provide height and colour at the back of the border, when the Delphiniums have finished. **H. *decapetalus Loddon Gold*** has vivid, deep yellow, double flowers in late summer. Good in clay; resents dry soil. Sun. Ht. 120cm. Sp. 40cm. Be warned this is a plant shamelessly glorying in its own lack of refinement. **H. *Lemon Queen*** has delightful medium sized pale lemon yellow flowers. Aug-Oct. It thrives in dry soil, and, despite being a sunflower, grows as well in shade as it does in sun. Ht. 200cm. Sp. 60cm. **H. *salicifolius*** is grown for its foliage. 250cm. stems clad in narrow drooping, willow like leaves. Looks wonderful with grasses. Much used by Piet Oudolf. Small yellow flowers in September. Runs gently. Makes a large, majestic clump with a spread of about 100cm.

HELIOTROPIUM. (Boraginaceae). Native to Peru, can in the wild grow to 200cm. tall. **H. *arborescens Marine*** bears sweetly scented deep violet flowers throughout the summer and the autumn. It grows best in a sandy loam. Feed with tomato fertiliser and water throughout the summer. Tender evergreen shrub. Easy to overwinter in a heated greenhouse. Victorian gardeners would often prune plants into standards for bedding out. Ht. 60cm. Sp. 50m.

HELLEBORUS. (Ranunculaceae). The name derives from the Greek words 'hellein' to kill and 'bora', food – a reference to the plant's poisonous properties. Hellebores light up the winter and early spring like no other plants. **H. *argutifolius* syn. H. corsicus** has pale green flowers in late winter, which are upturned rather than pendant like Helleborus foetidus. It has the best foliage of all the hellebores, thick, leathery, prickly and architectural. It needs an open position in well-drained soil and protection from icy winds. Ht. 60cm. Sp. 50cm. **H. *a. Silver Lace*** with bright shiny, silver leaves. **H. *cyclophyllus.*** Scented light greenish yellow flowers. Dec. Ht 40cm. Sp. 60cm.

Dahlia Pink Decorative *(5)*

Deciduous. Any well drained soil. Sun or part shade. **H. x *sternii*** is similar to H. argutifolius but has handsome deep red stems. Both the Latin and the English names of **H. *niger***, the Christmas Rose, are misleading. Niger refers to this Hellebore's black root rather than the flower, which is pure white and it seldom flowers at Christmas; it normally flowers From mid Jan through March. Ht. 30cm. Sp. 40cm. Sun or part shade. Hellebores are as rule very drought tolerant and will thrive in poor soil. The Christmas Rose will not. It needs rich, moist soil and will not put up with second best. Although the plant is perfectly hardy, its flowers are best protected with an open ended cloche from winter wet and frost. **H. x *nigristern***. A hybrid between H. niger and H. sternii, this is a super plant. Tough, vigorous and long lived. Large pink-flushed, white flowers which open flat and are held facing outwards the better for us to admire, not like the drooping flowers of H. niger, which seem bashful in displaying their beauty. Feb-May. Pewter flushed foliage and red stems. Ht. 45cm. Sp. 35cm. Sun or part shade. Any soil. I have listed H. foetidus and H. orientalis in our section for shade plants.

HERMODACTYLUS. (Iridaceae). **H. *tuberosa***. The Snake's Head Iris was an early introduction into this country. Gerard's description of it is irresistible. Please note when he uses the word "leaves", he means in this context petals. "The lower leaves that turne downward, are of a perfect black colour, soft and smooth as is black velvet, the blackness is welted about with greenish yellow, or as we terme it, a goose turd greene, of which colour the uppermost leaves do consist." E. A. Bowles description is also apt. "I love this weird little flower made up of the best imitation I have ever seen in vegetable tissues of dull green silk and black velvet – in fact it looks as if it had been plucked from the bonnet of some elderly lady of quiet tastes in headgear." The flowers are scented. March-April. Hot sunny spot. Well-drained soil. Ht. 30cm. Sp. 50cm. Spreading tubers. Plant it perhaps intermingled with Oenothera sp. Siskiyou, which enjoys similar conditions but flowers later.

HESPERIS. (Cruciferae). **H. *matronalis Alba***. Sweet Rocket has been grown in England since the beginning of the 16th Century. One of the six best cottage garden plants! Sweet smelling, white, phlox-like flowers May-July. Should be planted in great swathes. Ht. 60cm. Sp. 40cm. Superb with Oriental poppies. Thrives in poor dry, sandy soil. Lime tolerant. Sun or part shade. Short lived but self seeds. This plant was apparently a favourite flower of Marie Antoinette. At the end of her life when the queen was waiting in a squalid prison for the guillotine, the prison's concierge risked her life and brought to Marie Antoinette bunches of Sweet Rocket.

HIBISCUS (Malvaceae). **H. *manihot* syn. Abelmoschus manihot**. Palest sulphur yellow flowers with a purple eye. June-Oct. Ht. 150cm. Sp. 35cm. Tender but makes copious seed which is easy to germinate under gentle heat and flowers the same year from an April sowing.

HIERACIUM (Asteraceae). **H. *aurantiacum* syn. Pilosella aurantiaca**. Bears the odd but somehow appropriate English name "Grim The Collier" native to North Europe. Wonderful burnt orange flowers May-July. Hairy rich dark green leaves. Will grow in sun or part shade. Needs a little moisture. A thuggish coloniser, it is good ground cover for a not too dry bank. Ht. 40cm. Sp. 100cm. Jason Hill, after suggesting that you set this plant "in a weedy corner and watch it shoulder out the buttercups and wrestle with the docks," rightly cautions: "delicate plants have been known to wilt when they see it coming."

HUMULUS. (Urticaceae). **H. *lupulus Aureus***. The yellow leafed form of The Common Hop. Bright yellow leaves which in late summer slowly turn green. Dies back completely

Delphinium elatum *(3)*

in winter. Good for weaving through hedges. Ht. 200cm. Sp. 200cm. Not too dry soil. Sun or part shade.
HYSSOPUS. (Labiatae). **H. *officinalis.*** Hyssop has a long history as a medicinal herb and has been used in the treatment of asthma, bronchitis and rheumatism. Its essential oil is antiseptic. The foliage was used in the Middle Ages as an aromatic strewing herb. An easy evergreen subshrub, which thrives in sandy soil and full sun, it grows to about 45cm. in height and makes a spread of about 30cm. Its flowers, which are borne June-Aug, are much in favour with bees. It is sometimes used, as at Helmingham, for low hedging in knot gardens. Although not very long lived - about five years is its allotted span - it is easy enough to renew from cuttings or seed. The species has rich, deep blue flowers. **H. o. *Rosea*** has delicious, deep rosy pink flowers.
INCARVILLEA. (Bignoniaceae). **I. *delavayi.*** Bright, rose red trumpet shaped flowers June-Aug. Very hardy, despite its exotic looks. Ht. 50cm. Sp. 25cm. **I. *d. Alba*** has pure white flowers with a yellow throat - a most refined and elegant plant. June-Aug. Ht. 50cm. Sp. 25cm. Incarvilleas need rich, well drained soil in a sunny position. Watch out for slugs.
INULA. (Compositae). The name Inula is supposed to be a corruption of Helenula, meaning little Helen, a name it acquired because Helen of Troy was supposed to have been holding a bunch of Inula helenium when Paris abducted her to Troy. Inulas will grow happily in light or heavy soils. Like most Compositae, they need sun. **I. *magnifica*** has dark centred, yellow flowers. Shaggy petals. Statuesque border plant. Essential for its strong perpendicular lines. Bold basal rosette of large, mid green, heart shaped leaves. Good for dramatic flower arrangements. Its stems and seedheads look good in winter with frost on them. One of Piet Oudolf's favourites for planting with tall grasses. Ht. 150cm. Sp. 50cm. **I. *hookeri*** is a much more compact plant with bright green leaves and rich yellow, narrow petaled daisy flowers, which open from attractive furry buds. July-Sept. Ht. 50cm. Sp. 40cm.
IRIS. (Iridaceae). ***I. unguicularis* syn. Iris stylosa,** a name it got from the apparent stem of the flower being in fact the style, the ovary being at soil level. "Suppose a wicked uncle who wished to check your gardening zeal left you pots of money on condition you chose only one species of plant: what plant would you choose? I would settle on Iris unguicularis". So wrote E. A. Bowles, one of the few plantsmen to combine discrimination with wit. Iris unguicularis is a native of Algeria and has adapted itself to the harsh climate of that arid land. Resistant to the appalling drought of the North African summer, it flowers during the winter rains. Such a plant is ideally suited to my native East Anglia, thriving in our drought stricken sand. It does however require the protection of a south or west-facing wall, to ensure that it flowers reliably in our less mild winters. The flowers are best picked in bud, as even the slightest wind or rain will trammel their beauty: exquisite lavender flowers with the most delicate markings. It is one of the great pleasures of winter to seek the shy, tightly sheathed buds nestling among the drab rustling leaves, and carry them back indoors to sit by the fire and watch them unfold. Once planted, Iris unguicularis should be left undisturbed. It is a trouble free plant, apart from slugs, which sometimes regard their flower buds as an hors d'oeuvre before a main course of Hostas. Dec-March. Ht. 30cm. Sp. 30cm.
JOVIBARBA. (Crassulaceae). **J. *allionii.*** Used to be called Sempervivum allionii. Fresh green rosettes of Sempervivum like foliage. Trough or gravel garden. Open position and well-drained soil. Ht. 10cm. Sp. 25cm.

Dianthus barbatus *(3)*

KALANCHOE. (Crassulaceae). **K. *pumila*** is my favourite succulent. It makes a small rounded plant, covered in darkish blue, small succulent leaves. These are coated in a thick white farina. The apex of each leaf is toothed. In April-May the whole plant is covered in small rosy flowers. The overall effect is of complete refinement. Kalanchoe pumila comes from Madagascar and is tender. It is best kept under glass the whole year, as outdoors the farina will be washed from its foliage. Like all succulents it must be kept very dry in winter as over-watering at this time of the year will result in weak over-lush growth. It also should be grown in poor gritty soil; over-feeding will result in a loss of character. Kalanchoe pumila, like all succulents is generally free from parasites, although occasionally afflicted with mealy bug. Grows happily in a North-facing window. Ht. 20cm. Sp. 35cm.

KALIMERIS. (Asteraceae). Native to E. Asia, Kalimeris are graceful summer and autumn flowering daisy plants. They like sun and not too dry soil. **K. *incisa Charlotte*** grows 85cm. tall and bears blue flowers June-October. Sp. 50cm. **K. *i. Yomena Shogun*** is shorter, has variegated foliage and lighter blue flowers. Ht. 70cm. Sp. 50cm.

KITAIBELIA. (Malvaceae). **K. *vitifolia.*** Named after Paul Kitaibel, the Hungarian naturalist, this huge, herbaceous member of the mallow family came from Yugoslavia. Reliable and hardy. Graham Stuart Thomas is very rude about this plant, describing its flowers as gaping inanely at the skies. He is wrong. This is one of the very best back of border plants. Its large, vine shaped leaves always look fresh, even when starved of moisture. Its flowers are pure white and delicately incised, the green bracts being visible between each petal. July-Oct. Ht. 210cm. Sp. 80cm. Any well drained soil. Sun or part shade.

KNAUTIA. (Dipsacaceae). Named after the German 18th Century botanist Christian Knaut, Knautias are closely related to the Scabious family. **K. *arvensis***, the Field Scabious, bears bluish lilac flowers all summer. Any soil. Sun. Drought tolerant. 60cm. Ann Pratt, writes in her classic book on English Wildflowers: "This tall and handsome plant often overtops the ripening corn in June and July or is levelled with it by the reaper a month later." Ah, visions of a lost England! How often do we now see the Field Scabious growing in the cornfields? K. arvensis is strangely neglected as a garden plant. To my taste it is a far more beautiful plant than the Scabiosa caucasica hybrids. A large plant, growing to perhaps 120cm., it is much branched, making a spread of perhaps 80cm. It flowers prolifically from June through to October. It is an easy healthy plant, which, unlike the Scabiosa cauacasica hybrids, never seems to be afflicted with mildew. Geoffrey Grigson says girls used to pick Field Scabious "buttons", give each one a lover's name, and then choose her husband by the one which flowered best, hence another of its names, Bachelor's Buttons. **K. *macedonica*** used to be known as Scabiosa rumelica. It bears deep crimson flowers on tall wiry stems from May to September. It is one of those see through plants, which are good for planting at the front of the border. They create a fine tracery but do not obscure what is behind. Not very long lived but normally self seeds. To my mind unique and wonderful; once again I am in disagreement with Christopher Lloyd, who thinks its popularity "hard to justify". Ht. 90cm. Sp. 60cm. Both knautias need sun; they are happy growing in any well drained not too dry soil.

KNIPHOFIAS. (Xanthorrhoeaceae). Kniphofias get their name from the 18th Century German botanist J. H. Kniphof. They mostly originate in S. Africa. Kniphofias are easy plants in good moisture-retentive soil. They need good drainage and are liable to rot in

Dianthus Lady Acland Pink *(3)*

badly drained clay. I am not very fond of the classic old fashioned red and yellow pokers. Red and yellow to my mind always look dirty commingled on the same flower, although separately they have great zest. However the new hybrids with their much more subtle colouring I think are quite beautiful. **K. *Bees Lemon.*** Dense heads of green buds which open lemon yellow. June-Sept. Ht. 85cm. **K. *Brimstone*** has slender, green and pale yellow pokers. Aug-Dec. Ht. 75cm. Sp. 35cm. **K. *caulescens*.** Brownish orange flowers, which fade to cream. June-Aug. Ht. 100cm. Sp. 40cm. **K. *Dingaan*.** Bronzed yellow flowers. A seedling from uvaria nobilis. The tallest yellow poker. Ht. 150cm. July-Aug. **K. *Dorset Sentry.*** Fat pokers of clear, acid yellow flowers July-Oct. Ht. 100cm. Sp. 40cm. **K. *Jenny Bloom*** has narrow heads of delicate coral buds, which open to creamy flowers. Choice and very unthreatening. June-Oct. Ht. 70cm. Sp. 35cm. **K. *John Benary*** Old Irish Hybrid. Coral red flowers July-Sept. Ht. 75cm. **K. *Light Of The World.*** Lovely soft orange flowers held in slender heads. Aug-Oct. Ht. 60cm. Sp. 35cm. **K. *Little Maid*** is the dwarfest of its tribe. It has the softest, creamy yellow flowers, which are produced in abundance from June to November. It makes nice small tussocks of grassy foliage. Introduced by Beth Chatto. Ht. 50cm. Sp. 35cm. **K. *Nancy's Red*** has slender heads of dusky, deep coral pink red flowers. Utterly different and distinctive. July-October. Ht. 60cm. Sp. 35cm. **K. *Percy's Pride*** makes statuesque mounds of foliage from which rise imposing spikes of greenish flowers. Quite wonderful. Percy is deservedly proud! August-November. Ht. 75cm. Sp. 50cm. It looks superb with Dahlia The Bishop of Llandaff. **K. *rooperi*.** Late flowering poker with large fat heads of orange and yellow flowers. Plant with Miscanthus and Molinias. Sept-Nov. Ht. 130cm. **K. *Tawny King*** has tawny apricot buds which turn cream when opened. Long flowering. July-Oct. Bob Brown of Cotswold Garden Flowers says this is their most asked for Kniphofia on his nursery. Ht. 120cm. Sp. 40cm. **K. *Timothy*.** Soft salmon, peach and cream flowers. Bronze stems July-Sept. Ht. 70cm. Sp. 40cm. **K. *uvaria Nobilis.*** Bold spikes of orange flowers Aug-Sept. Ht. 200-400cm. Sp. 100cm. Wonderful in the middle of a group of salvia uliginosa. **K. *Wol's Red Seedling*.** Narrow red flowers. Floriferous. July-Aug. Ht. 60cm. Sp. 40cm.

LACTUCA. (Compositae). **L. *perennis*.** Violet daisy flowers May-July. Finely cut deep green leaves. Ht. 25cm. Sp. 30cm. Disappears soon after flowering, leaving space for such late appearers as Colchicums and Nerines.

LATHYRUS. (Leguminosae). Lathyrus is the ancient Greek name for pea. **L. *vernus*** was introduced from Europe before 1629. It was grown by Parkinson under the name of the "Blew Everlasting Pea". Purple flowers in March-May. Much divided foliage. Compact. Ht. 40cm. Sp. 30cm. Sun. Happy in any soil that does not become waterlogged. A very pretty, dainty, spring plant. Even prettier is **L. *v alboroseus.*** Identical in habit to the species plant its flowers are a delightful pink and white confection. **L. *latifolius*** the perennial Sweet Pea was grown by Gerard in his garden at Holborn. **L. *l. White Pearl*** has white flowers July-Sept. An herbaceous plant, it climbs each year to 150cm. Summer flowering. Cut to the ground each autumn. Useful for growing through early flowering shrubs. Once established, totally drought tolerant, its roots go deep as a vine's. Lathyrus latifolius also rival vines in longevity, I know one for certain which is over fifty years old. **L. *sativus*.** A Mediterranean annual but well worth the trouble. Exhillarating sky blue flowers. June-July. Ht. 100cm. Sp. 25cm. Easy.

LAVANDULA. (Asteraceae). Lavender gets its name from the Latin for to wash, due to the universal ancient Roman custom of abluting with lavender water. Lavenders are

Dodecatheon meadia *(3)*

Mediterranean plants and, although usually found in the wild on limestone, they will thrive in any hot, well-drained soil. Like many Mediterranean aromatic shrubs, they have a limited life and after 10 years or so become woody and gaunt and will need replacing. Lavenders should have the flowering stalks removed in autumn, but must not be pruned till new growth has begun in the spring, when they should be cut back hard, removing all the previous season's growth. **Hardy Varieties.** These are all hardy except in the heaviest wettest soil. All ***L. angustifolia*** varieties flower July-Sept. All have very narrow silver leaves and make tight growing plants suitable for dwarf hedging apart from Twickel Purple. All have a spread of about 35cm. and flower from July to September. **L. *a. Arctic Snow*.** White flowered lavender with narrow silver foliage. Ht 45cm. **L. *a. Ashdown Forest*** has narrow green leaves and clear light blue flowers. Strong grower. Ht. 60cm. **L. *a. Hidcote*.** Very deep blue flowers, narrow silver foliage and a compact habit. Always a favourite for miniature hedging. Ht. 45cm. **L. *a. Imperial Gem*.** Similar to Hidcote in flower colour but shorter and healthier. Ht. 40cm. **L. *a. Lavenite Petite*.** Very early flowering. Dark blue pom-poms. Compact. Ht. 35cm. **L *a. Little Lady*.** Clear light blue flowers. Free flowering. Very compact. Ht. 40cm. **L. *a. Miss Katherine*.** Pink flowers. Ht. 60cm. According to Christopher Fairweather the best pink flowered lavender. **L *a. Munstead*** has lavender blue flowers and broadish silver leaves. Compact. Looks better in winter than L. Hidcote. Ht. 45cm. Sp. 40cm. **L. *a. Twickel Purple*** has long dark blue flowerheads with spaces between the whorls. Ht. 75cm. **L. x intermedia and L. vera.** Varieties in this group are stronger growing than L. angustifolia and are suitable for use as large specimen plants or for massing with cistuses and other bold Mediterranean shrubs. L. x intermedia's wide silver leaves remain presentable in winter; all varieties flower July-Sept. and have a height of about 90cm. With a spread of at least 60cm. **L. *x intermedia Fragrant Memories*** has medium purple flowers. **L. x *i. Sawyers*** has particularly intense silver foliage and rich purple flowers – slow to get established but worth the wait. **L. *vera Walberton Silver Edge*.** Silver edged leaves. Lavender violet flowers. July-Sept. Strong growing plant Ht. 70cm. Sp. 50cm. **Lavandula stoechas.** The following varieties of L. stoechas are less reliably hardy than L. angustifolia and L. x intermedia and need a sheltered warm position. They flower July-Sept and have a spread of about 35cm. **L. *s. Pretty Polly*.** Small blue flowers topped with white flags. Compact 40cm. **L. *s. Regal Splendour*.** Purple violet flags and royal purple flowerheads. Compact and aromatic. Ht. 70cm. **Tender Varieties.** These varieties make handsome potplants and if moved into a warm conservatory in winter will flower throughout the year. **L. *Kew Red*** is a personal favourite. It has very fine green foliage and dark pink flowers. Ht. 40cm. **L. *pinnata*** has filigree silver foliage and lovely candelabra heads of violet blue flowers. Ht. 50cm. Sp. 45cm.

LAVATERA. (Malvaceae). Lavatera was named after Lavater, the 16th Century Swiss naturalist. A shrubby member of the mallow family, it thrives on the light sandy soils of East Anglia. **L. *Barnsley Pink*** is a very vigorous hardy shrub with white flowers with a pink blush. June-Oct. Grows to 250cm. Sp. 250.cm. **L. *Burgundy Wine*.** Deep red flowers. Compact habit. Ht. 100cm. Sp. 80cm. **L. *Icecool*** has white flowers fading to soft pink. June-Sept. Ht. 150cm. Sp. 80cm. Best treated as a herbaceous plant and cut to the ground in the spring. **L. *White Angel*** bears pure white flowers June-Sept. Ht. 150cm. Sp. 80cm. **L *maritima*** is a refined plant. Its flowers, which have reflexed petals and a very protuberant stigma, are more suggestive of a Hibiscus flower than Lavateras such as Olba Rosea and Barnsley Pink. The petals are of a soft mauve, which

Echinacea purpurea *(3)*

deepen in colour to rich purple in the centre of the flowers. June-Nov. Its foliage is slightly glaucous and vine shaped. Lavatera maritima normally grows to not more than 140cm. height and 120cm. in spread and is, therefore, easier to accommodate than some of its more boisterous sisters. It is, however none too hardy and in all except the warmest gardens will need the protection of a west or south facing wall. Given such a favoured position and well-drained soil, it will thrive and prove as floriferous as Barnsley Pink. Gardeners who get the autumn pruning itch are politely asked to keep their secateurs under control, because Lavatera Maritima, like so many half-hardy shrubs, resents the autumnal massacre and should be pruned when in growth in late spring.

LEPTINELLA. (Asteraceae). L. ***squalida Platt's Black*** has evergreen black ferny prostrate foliage. Good ground cover. Sometimes used for small lawns. Spreads stoloniferously. Needs moist but well drained soil. Ht. 5cm.

LEUCANTHEMUM. (Asteraceae). Leucanthemums get their name from the Greek word for white, 'leukos', and 'anthemon', flower. Native to Europe and N. Asia, they are extremely reliable, hardy plants. **L. *superbum Becky***, a selected form of the single white Shasta daisy, is unbeatable for its huge, perfectly formed, white daisies on strong 90cm. stems - ideal for cutting. Sp. 40cm. Rich dark green foliage. A wonderful plant untroubled by pest or disease. Happy in any soil. June-August. **L. *s. Sonnenschein*** has large, creamy yellow, single flowers, which pale with age. July-August. Ht. 70cm. Sp. 40cm. All Shasta daisies need to be frequently dug up and divided in order to maintain their vigour.

LEUCOJUM. (Amaryllidaceae). The Greek name Leucoion means a white violet. Its English name is "spring Snowflake". Gerard grew **L. *vernum*** in his garden at Holborn. White flowers with green tips in Feb-April. Ht. 15cm. Sp. 25cm. Does best in rich moist soil. Part or full shade.

LEWISIA. (Portulaceae). **L. *cotyledon Alba.*** I am no lover of the bright sunset colours of most lewisias but the pure white flowers of this variety are irreproachable. June-Aug. Ht. 20cm. Sp. 25cm. Lewisias hate winter wet on their evergreen rosettes and should be planted on their side in a not too hot wall.

LIATRIS. (Asteraceae). **L. *spicata Alba.*** A very unlikely looking member of the daisy family. White bottlebrush flowers July-August. Ht. 75cm. Sp. 40cm. **L. *spicata Kobold.*** is a compact form with violet lilac flower spikes. Ht. 50cm. Sp. 40cm. Liatris require moist fertile, but well drained soil; if the ground is waterlogged in winter their tubers will rot. If cut for flower arranging it is important not to cut the flowering stems to the ground, as development of the tubers is dependant on the foliage maturing.

LILIUM. (Liliaceae). The name lily comes from the Greek word 'leirion', used by Theophrastus for the Madonna Lily. **L. *candidum*.** Possibly the oldest cultivated plant. It is represented on Cretan vases of the Minoan period (1750-1600 B.C.). The first record of it in England is its appearance in a 10th Century miniature of Queen Ethelreda, the founder of Ely Cathedral. The Madonna Lily is not the easiest of plants. It likes a rich alkaline soil and full sun. Do not crowd it if you want blooms fit for the hand of the archangel Gabriel. Ht. 130cm. Sp. 30cm. **L. *formosanum Pricei*** is a delightful miniature species lily with large trumpet shaped ivory flowers striped with dark red on the outside. June. Ht. 15cm. Sp. 25cm. A fanfare of these can almost redeem that heap of rubble you call a rockery. **L. *Casablanca*** has scented pure white flowers with reflexed petals - lovely in pots on the terrace. Ht. 80cm. Sp. 35cm. **L. *White American*** has scented,

Echinops sphaerocephalus *(3)*

pristine white trumpets. Clean rich green leaves. June-July. Ht. 60cm. Sp. 25cm.
LIMONIUM. (Plumbaginaceae). **L. *latifolium* syn. L. platyphyllum** is a perennial form of the statice or sea lavender, native to South Eastern Europe. The word Limonium is derived from the Greek word, leimon, meaning marsh or meadow, which is the preferred habit for the common Sea Lavender, Limonium vulgare. Salt tolerant. Confusingly, Limonium latifolium, which is tap rooted, thrives in poor dry soil. It makes a basal rosette of dark green leathery leaves and throws up 60cm. stems of lavender flowers. July-Sept. It is one of those airy plants, which, like Gypsophylla or Crambe, are wonderful for lightening the border. It is also a good plant for the gravel garden. **L. *chilwellii*** has white buds which open into violet flowers. Branching and upright. Aug-Sept. Ht. 60cm. Sp. 40cm. This is supposed to be a better more vigorous plant than L. Violetta. Christopher Lloyd says Limoniums in all their parts have an ill-favoured fragrance.
LINARIA. (Plantaginaceae). All linarias thrive in full sun and well drained soil. Tolerant of drought, they dislike heavy clay. **L. *purpurea.*** A native of Southern Italy L. purpurea was introduced into English gardens in the early 17th Century. This tall Toadflax has always epitomised to me the softness of the traditional English garden. Its light airy spires are useful for providing uplift at the front of the border in late spring and early summer, performing much the same role as Verbena bonariensis does in late summer and early autumn. It also makes a super cut flower - lovely with old fashioned roses. L. purpurea is a welcome and reliable self-seeder. The pink and white forms reproduce about 45% true from seed. **L. *purpurea*** has purple flowers, **L. *p. Canon Went*** has the most delicious shell pink flowers, and the very rare **L. *p. Alba*** has pure white flowers. All have a height of 75cm. and a spread of about 30cm. They flower June-Sept.
LINUM. (Linaceae). **L. *perenne*** comes from North West America. It was introduced into cultivation in the 16th Century. It grows best in a good loam and needs full sun. It bears brilliant blue flowers throughout the summer. June-Aug. Ht. 45cm. Sp. 25cm.
LITHODORA. (Boraginaceae). **L. *diffusa Heavenly Blue.*** Native to South West Europe. Gentian blue flowers. May-July. Rich semi evergreen foliage. Prostrate habit. Needs light acid soil. Ht. 15cm. Sp. 30cm. Looks wonderful with Helianthemum Henfield Brilliant. One of the three best blue flowered plants! The other two being S. uliginosa and Omphalodes Cherry Ingram.
LOBELIA. (Campanulaceae). Mention the word lobelia and most people think of shapeless bedding plants, not the soaring spires of L. cardinalis, L. syphilitica, and their hybrids. These latter are essential plants for the late summer and autumn garden; they provide the strong, perpendicular lines, which plants like foxgloves and mulleins contribute earlier in the season. They are easy plants for the moist border or bog. Those with sandy soil need to dig in plentiful well rotted humus. A thick protective mulch over the resting buds is advisable in winter. Lobelias clump up well and should be divided every second year in May. If left undivided, they tend to rot. They all grow to about 90cm. **Dark leafed varieties. L. *fulgens Elmfeur*** (Introduced from Mexico by Humboldt and Bonpland in 1810.) Has rich, dark red foliage and bright red flowers; the superb clash of its foliage and flowers make it an essential ingredient for the "hot border". Unlike L. cardinalis Queen Victoria, which it closely resembles, it is both reliably hardy and tolerates dry soil. July-October. Ht. 90cm. Sp. 20cm. **L. *Hadspen's Purple.*** Vibrant purple flowers above dark green foliage. July-Oct. Ht. 90cm. Sp. 30cm. **L. *Ruby Slippers.*** Ruby garnet flowers. July-Ot. Ht. 120cm. Sp.

Eryngium alpinum *(1)*

30cm. L. ***Russian Princess*** has magenta flowers and bronze foliage - a sumptuous combination. July-Dec. Ht. 90cm. Sp. 20cm. **Green leafed varieties. L. *Tania*** has crimson purple flowers. July-September. Ht. 110cm. Sp. 20cm. **L. *x vedrariensis*** has spikes of dark blue violet flowers, which look good with Heleniums. July-Oct. Ht. 90cm. Sp. 20cm. Very different from the above plants is **L. *tupa*.** A native of Chile, it makes a spreading plant with large matt sage coloured leaves, crowned by beautifully articulated brick red flowers. It needs rich moist soil. July-December. Ht. 180cm. Sp. 100cm.

LUPINUS. (Leguminosae) Russell Lupins are something of a rarity in the modern garden, resented for their girth and ephemeral beauty, ousted in favour of neat conifers and concrete. I am a sucker for all things ephemeral. I love flowers which only flower briefly. If something is there the whole time, we cease to be surprised by it and lose the gift of looking at it afresh. Lupins in June are majestic with their spreading branches, all decked in the freshest of green and their scented flowers like miniature horse chestnut blossoms. They are undemanding plants, which grow well in any well drained soil, which is not too alkaline. All they demand is sun and space. All the following flower May-July and have a height of 100cm. and a spread of 60cm. **L. *Noble Maiden*** has cream flowers. **L. *The Governor*** has bicoloured blue and white flowers. **L. *The Chatelaine*** is a pink and white bicolour. **L. *The Pages*** has carmine flowers. For those with plenty of space grow the Tree Lupin, **L. *arboreus***, (introduced by Captain Vancouver from California in 1792). This is available in three colours: **L. *a. Yellow Form***, **L. *a. Blue Form,*** and **L. *a. White Form*.** These are true shrubs and should be spared the secateurs and allowed to grow into dense 200cm. bushes. Their spread will be at least 100cm. They flower May-Sept. Lupins are impervious to drought, once their taproots are established, but must be watered in the first season after planting. Lupins need good drainage and will not overwinter on heavy clay. The tree Lupins are more tender than the Russell Lupins and are not suitable for very cold areas.

LYCHNIS. (Caryophyllaceae). **L. *chalcedonica*.** The Maltese Cross Plant. Russian in origin, it appears that L. Chalcedonica was cultivated by the Byzantines in their gardens in Constantinople, from whence it was brought back to Europe by the crusaders. Lychnis chalcedonica bears true scarlet flowers in June-July. It does well in heavier soils but its flowering stems are rather weak and need some support if they are not to be flattened by rain. Ht. 80cm. Sp. 35cm. C. Lloyd recommends planting it with Salvia superba. **L. *coronaria*** (The Rose campion). This, Alice Coats tells us (with no source quoted) is said to have sprung from the bath of Aphrodite. The Greeks used the grey felted leaves of Lychnis coronaria as wicks in oil lamps; the Greek word for lamp is "lychnos". The Rose campion was in cultivation in England by the early 14th Century. L. coronaria has carmine red flowers and **L. *c. Alba*** has pure white flowers. They are both heavily drought resistant and do well in dry sandy soil. July-Aug. Ht. 40cm. Sp. 30cm. Lovely with the blue chicory, Cichorium intybus. Lychnis coronaria is one of those plants which we are over-familiar with and need to look at with fresh eyes. The snobbish preoccupation of the fashionable gardener with modern rarities at the expense of old beauties is nothing new. Shirley Gibberd writing in 1898 had obviously experienced her share of 19th Century plant snobs: "In the spring season the huckster florists sell rose campions at a penny each....whoever would despise them for their cheapness would deserve to see no other flowers in this world whatever he might see in the next." **L. *flos-cuculi*.** The native Ragged Robin, or Shaggy Jacks as it is called in

Filipendula rubra Venusta *(3)*

Devon, is so called for its flowers "being finely and curiously snipped in the edges" (Gerard's Herbal). With its narrow twisted petals and brown, bladder shaped calyxes, it demands botanical illustration. Robert Gathorne Hardy wrote of this plant "no abundance can stale". It likes moist soil and flowers May-June. The pure white form **L. *f. Alba*** is, perhaps, my favourite white flowered plant. Both forms are about 40cm. high with a spread of about 25cm. Although in the wild they are always to be found in boggy ground they are happy in any except the driest soil. **L. *viscaria***, the Sticky Catchfly although native is rare to Britain. It derives its name from the sticky patches on the stems just below each leaf junction. It likes dry soil and good drainage. **L. *v. Alba*,** a pure white form May-June, Ht. 40cm. Sp. 20cm. and **L. *v. Splendens*** has carmine pink flowers June-July, Ht. 50cm, Sp. 20cm.

MALVASTRUM. (Malvaceae). **M. *lateritum*.** Perhaps my favourite member of the mallow family. Small deep green leaves. Elegant apricot flowers with an inner red ring. June-Sept. Needs a rich soil and a warm spot. Grows spectacularly at Sissinghurst against a sunny wall, which enables it to retain its old growth over winter. Grown in the open border it tends to get cut to the ground by the frost each year and not make much of a show. Ht. 30cm. Sp. 80cm.

MATTHIOLA. (Brassicaceae). **M. *fruticulosa Alba*.** The White Perennial Stock. Grows wild on the Isle of Wight. A splendid creature, hardy and perennial, with very silver foliage, tending to make a large sprawling plant. The main flush of flowers is produced in May, when to walk past a plant in evening is to experience instant, nasal intoxication. If deadheaded, it will continue to produce flowers in lesser profusion till September. Matthiola plants tend to become woody and unsightly after a few years, when they should be pulled up to make way for younger plants, of which you should have a plentiful supply, as they are generous self-seeders. These stocks are happy in the poorest dry sandy soil and thrive in seaside gardens. They like full sun and will not abide heavy clay or the unremitting cold experienced in frost pockets. May-September. Ht. 60cm. Sp. 40cm.

MELIANTHUS. (Melianthaceae). **M. *major*.** Perhaps the most magnificent of all foliage plants, this South African giant needs a warm sheltered spot and rich fertile soil. Grey green deeply cut leaves, which when bruised give off a sour sexy scent. In mild winters evergreen. If not cut back by the frost in winter, it is a good idea to give it a sharp prune in spring. Unpruned it quickly becomes gangly. Melianthus major, left unpruned, will flower in May, but its small maroon flowers are of little significance - "Once seen best forgotten" says Christopher Lloyd. Melianthus is stoloniferous and needs an area at least 250cm. in diameter. Ht. 200cm. Good in large tubs. Christopher Lloyd rightly says Melianthus "is as beautiful a foliage plant you can have in your garden".

MIMULUS. (Plantaginaceae). Mimulus probably gets its name from 'mimo', the Latin word for ape, a reference to the gaping mouth of the corolla. **Herbaceous varieties.** **M. *bartonianus*** has dusky deep pink flowers. July-Aug. Ht. 90cm. Sp. 30cm. **M. *cardinalis***, a native to the west coast of America was introduced into England in 1835. It is a hardy robust plant, which grows to 80cm. Ht. Sp. 30cm. It has clear scarlet flowers and fresh green foliage. It flowers throughout the summer and responds well to harsh curtailing after each flush of flowers. Alice Coats in her fascinating book 'Flowers and Their Histories' describes a peculiarity of this plant. M. Cardinalis, she writes "has very sensitive flowers; if the stamens are touched with pollen of its own kind, the flowers will close, and remain closed; but if dusted with sand, pollen from another species, dead pollen from its own, or other foreign body, it will close for a time, but will reopen later."

Fritillaria imperialis *(3)*

M. cardinalis thrives in most soils and does not need to be especially moist. **M. *cupreus Red Emperor*** is a compact prostrate plant. Rich red flowers in profusion. July-Sept. Ht. 15cm. Sp. 30cm. Spreading. **Shrubby varieties.** These varieties all come from Mexico and make small shrubs with attractive, sticky, rich green foliage. They are best against a sunny wall in rich, well drained soil. They make handsome pot plants for the terrace in summer, flowering incessantly well into November. They need frequent stopping, as they are apt to become straggly if allowed to grow unchecked. If grown in pots, they should be repotted regularly to keep them looking fresh. They all flower from June to November and grow to about 60cm. with a spread of about 40cm. **M. *aurantiacus*** has apricot coloured flowers. **M. *a. Popacatapetl*** has the most elegant white flowers. **M. *a. var Puniceus*** has rich dark red flowers. **M. *bifidus Verity Buff*** has creamy flowers.
MONARDA. (Lamiaceae). Monarda is named after the Spanish 16th Century botanist Nicholas Monardes. Its English name is Bergamot. All varieties have strongly aromatic foliage. Monardas are moisture loving plants and, in poor dry soil, every leaf and petal will express utter misery and reproach. In good rich soil they are one of the great flowering plants of summer. **M. *Fishes*** has the clearest, pale pink flowers. July-August. Ht. 90cm. Sp. 40cm. **M. *Ruby Glow*** has dark, ruby red flowers over a long season. June-Sept. Red flushed foliage. Ht. 75cm. Sp. 30cm. Sun. Moist rich soil. **M. *Snow Queen*** has pure white flowers. July-August. Ht. 100cm. Sp. 40cm. **M. *Squaw*** has bright red flowers. July-August. Ht. 100cm. Sp. 40cm.
MORINA. (Morinaceae). **M. *longifolia.*** Prickly, dark green rosettes. Leaves have spicy lemon scent when bruised. Whorls of white flowers which age to pink. Resembles a sylph-like Acanthus. One of the six most elegant plants in my garden! Needs moist rich soil and an open position. Self seeds. June-Aug. Ht. 75cm. Sp. 50cm.
MYRTUS. (Myrtaceae). **M. *communis.*** A native of the Mediterranean, in this country needs the protection of a warm south or west wall. In very harsh winters it will still get cut by the frost, but always rejuvenates itself. Lime tolerant. Wonderful aromatic foliage and creamy flowers, traditionally used in Mediterranean bridal wreaths and bouquets. Glossy, deep green foliage. "Virgil describes the Myrtles as *amantes littora myrtos*, and those who have seen the Myrtle as it grows on the Devonshire and Cornish coasts will recognise the truth of his description". (Canon Ellacombe). **M. *c. Tarrantina*** has narrow, almost needle like, deep green leaves. It flowers July-Sept. Ht. 200cm. Sp. 150cm.
NEMESIA. (Scrophulariaceae). **N. *denticulata.*** Most Nemesias are tender and grown as bedding plants. These appear to be hardy. They have fresh green foliage and cover themselves with flowers June-Dec. Nemesias, related to diascias, which I heartily detest for their messiness, are elegant, airy creatures. They need sun and well drained soil. Drought tolerant. **N. *d. Confetti*** has pale pink flowers Ht. 35cm. Sp. 35cm. **N. *d. Golden Eye*,** has dark purple flowers with a bright golden eye. Patrick Fairweather reckons it's a stunner. Perhaps not quite as hardy as N. Confetti. Very compact habit. Ht. 20cm. Sp. 20cm
NEPETA. (Lamiaceae). The Latin name Nepeta seems to derive from the place name Nepi in Italy. The English name catmint derives from cats' passion for some forms of this plant. J. P Tournefort in his book 'The Complete Herball, or The Botanical Institutions of Mr Tournefort', published in 1730, tells us that Nepeta was credited with the property of enbravening the timid. He relates how a hangman "that was otherwise gentle and pusillanimous never had the courage to behead or hang anyone, till he had first chewed the root of the Catmint". **Drought loving Nepetas.** All the drought loving

Fritillaria imperialis *(6)*

Nepetas have aromatic foliage. They all respond well to a mid season assault with the shears, flushing up with fresh foliage within days. **N. *Walkers Low,*** a superb variety, with rich deep blue flowers and a good upright habit. May-July. Ht. 60cm. Sp. 50cm. **N. *grandiflora Dawn to Dusk*** has smoky pink flowers with darker bracts. July-Sept. - looks very effective with Campanula rotundifolia. Ht. 90cm. Sp. 40cm. **Moisture loving Nepetas.** These Nepetas need rich moist soil and will tolerate light shade. **N. *govaniana*** is native to the Western Himalayas. It bears erect 120cm. spikes of soft yellow flowers July-Sept. It is the most elegant of all the Nepetas. Its foliage has an attractive lemony scent. **N. *sibirica Souvenir d'Andre Chaudron.*** Blue flowers, deceitfully like a Penstemon's. July-Aug. My subjective nose says the aromatic leaves give off a provocatively erotic odour. Ht. 75cm. Sp. 40cm.

NERINE. (Amaryllidaceae). **N. *bowdenii.*** Discovered by a Government Surveyor with the splendid name of Mr Athelstan Cornish Bowden growing near King William's town in S. Africa in 1889.Vivid pink flowers Oct-Nov. Ht. 60cm. Sp. 20cm. A wonderful plant on our heathland sand. It used to be common to see hereabouts old cottages flaunting against a southern wall, a whole bed of them, packed tight. Swept away by gravel chips and carports. Nerines look lovely with silver artemisias, the chocolate leafed Geranium Dusky Crug and Salvia microphylla.

NICOTIANA. (Solanaceae). Nicotiana is named after the 16th Century French consul in Portugal, who introduced the plant to France. Tobacco plants have great grace and elegance and provide flowers into late autumn. They like rich, moisture retentive soil and thrive in part shade. **N. *sylvestris*** is half hardy and, although cut by frosts, often shoots from the base in spring. **N. *langsdorfii*** is tender, but easily propagated from seed. **N. *sylvestris*** is one of the most majestic plants for the border. It has huge green leaves and enormous sprays of white flowers, which waft an overwhelmingly sensual scent in the evening. July-September. Ht. 80cm. Sp. 40cm. **N. langsdorfii** has drooping sprays of greenish yellow flowers. July-September. Ht. 75cm. Sp. 30cm. It is much sought after by ladies of good taste, who remain inviolate to strong colour.

NIGELLA. (Ranunculaceae). **N. *damascena.*** "Love In The Mist" was first introduced into England in about 1570. The Latin name Nigella is a reference to its black seeds. In the 18th Century with its passion for regularity this fairy beauty was considered a curiosity rather than a beauty. Fairchild wrote in 1722 it is "rather an odd Plant than beautiful in its Flower; for the Blossom is of a very pale blue Colour and is encompass'd with shagged Leaves, as if it was ty'd up in a Bunch of Fennel". Love in the Mist is a Mediterranaean annual but because of its happy self seeding habit once introduced into the garden becomes a permanent feature. Ht. 45cm. Sp. 25cm. June-July. **N. *d. Miss Jekyll*** is an improved form.

OENOTHERA. (Onagraceae). Evening primroses require full sun and good drainage. **O. *speciosa*** is a mouth watering plant. Although only about 30cm. in height, it has flowers, which are as big as our native Oenothera biennis. But instead of a harsh, dirty chrome yellow, it bears the prettiest white flowers, heavily blushed and veined with pink. The flowers remain open throughout the day. Native to Texas, O. speciosa is stone hardy in our heathland soil. Though somewhat late to emerge from the winter recess, this plant makes up for lost time by flowering from mid June right through to the frosts. Its flowers are unscented. **O. *s. Siskyou*** is a much improved selection, which makes a neat defined clump with strong upright flowering stems, much in contrast to the species plant, which tends to sprawl like one of my adolescent sons across a sofa. Sp. 35cm.

Gentiana acaulis *(3)*

ONOPORDUM. (Asteraceae). O. ***acanthium.*** I am told the word onopordum in Greek signifies a donkey breaking wind. Before being submitted to a donkey's digestive juices, it is a great sculptural monster. 200cm. of twisting white branches, terminating in rich purple flowers. August-September. Looks superb against a barren landscape, such as gravel or parched grass. I sometimes imagine them like so many Don Quixotes in their steely armour, charging across the dunes at Sizewell, jousting with those castles of arrogance. Sp. 90cm.

ORIGANUM. (Lamiaceae). Origanum takes its name from the Greek words for mountain – 'oros', and joy – 'ganos'. It is a true Mediterranean plant, which thrives in hot dry conditions and flowers from late summer through into autumn. All varieties make dense mats of leaves and look good in gravel. They are much loved by bees. **O. *laevigatum Herrenhausen*** has rich purple flowers. July-Sept. Ht. 30cm. Sp. 40cm. **O. *rotundifolium Pagoda Bells*** has large heads of pink pendant bracts – a bit like O. dictamnus but easier to grow. July-Oct. Ht. 20cm. Sp. 30cm. **O. *vulgare Country Cream*** has variegated foliage and pale pink flowers. June-July. Ht. 15cm. Sp. 25cm. The foliage of O. v. Country cream is highly aromatic and is excellent in tomato sauce. **O. *v. Pink Mist*** has shocking pink flowers on dark stems. June-July. Intense yellow leaves. Ht. 20cm. Sp. 30cm.

OSTEOSPERMUM. (Asteraceae). **O. *jucundum Weetwood.*** I am no great lover of Osteospermums, but this one is different. It has a dense compact habit, rich green foliage and bears masses of pure white flowers with jet black centres. June-Oct. It is stone hardy. Ht. 20cm. Sp. 40cm.

PAEONIA. (Paeoniaceae). Paeonies are named after the mythical physician to the gods, Paion, who used its roots to cure a wound given to Pluto by Hercules. The Ancient Greeks believed the roots of paeonies "must be uprooted by a hungry dog tied to it by a string and enticed with the smell of roasted meat, the groan of the plant as it left the ground being fatal to all who heard it". (Alice Coats, 'Flowers And Their Histories'). What a splendid subject for a cartoon! P. officinalis, native to the Mediterranean was introduced to England in the mid 16th Century, P. peregrina from the Balkans in the early seventeenth century and P. lactiflora in about 1784 by the Russian traveller Pallas who found it growing in Siberia. The Mongolians are said to have used the roots of Paeonia lactiflora as an ingredient for soups and to have infused ground paeony seeds with their tea. Herbaceous Paeonies are essential cottage garden plants. They are decorative in all their parts. The young leaf shoots, which show through the soil in January are rich red and make a lovely backdrop for Snowdrops and Winter Aconites. Their adult foliage is handsomely cut and turns vivid red in autumn. Their seedheads, which are held high, split open to reveal rich red or blue seeds. And the flowers, whether ruffled and doubled, or single and cup shaped are the most sumptuous and voluptuous of all garden plants. Paeonies are extremely long lived, if not eternal, in the right conditions. They need a free circulation of air; in dank and stagnant gardens they will become prone to rot. When planting allow them a space at least 45cm. in diameter. They need well drained soil; like many rhizomes, while enjoying moisture in spring, they need a drier rest period in the summer. They should not be overfed; a light dressing in winter with some slow release fertiliser should be sufficient. They are happy either in sun or light shade. They must not be planted too deep; their crowns should be just below the surface of the soil. Deep planting is the most common cause of Paeonies refusing to flower. Young paeony plants often take a couple of years to establish themselves and begin their long flowering life.

Gentiana septemfida *(4)*

P. *lactiflora. **Buckeye Belle*** has semi double dark maroon flowers May-June. Ht. 85cm. **P. *l. Bowl of Beauty*** has single, pink flowers filled with cream petaloids. June-July. Ht. 100cm. **P. *l. Claire de Lune*** has ivory yellow flowers set off by orange anthers. May-June. Ht. 80cm.**P. *l. Edens Perfume*.** Large double pink flowers. Very free flowering. Ht. 70cm. **P. *l. Festiva Maxima*** has large fragrant round double white blooms flecked with crimson. June-July. 100cm. **P. *x l. Flame*.** Hybrid between P. lactiflora and P. peregrina. (Glascock 1939) Single brilliant red flowers. May-June. Ht. 75cm. **P. *l. Immaculee*** has pure white double flowers. Ht. 90cm. **P. *l. Jan van Leeuwen*.** Japanese form. Single white flowers. Late flowering. Ht. 75cm. **P. *l. Kansas*** has deep carmine red double flowers. Very reliable. May-June. Ht. 100cm. **P *l. Krinkled White*** has large single white flowers. Copes well with drought. Long lasting as a cut flower. Better I think than P. White Wings. **P. *l. Lady Alexandra Duff*** bears abundant small blush pink flowers. June-July. Ht. 100cm. **P. x. *l. Little Medicineman*.** Single pink flowers. May-June. Very compact. 70cm. Good in pots. **P *l. Monsieur Jules Elie*** has fragrant rose pink double flowers. Free flowering. The most popular pink variety in the U. S. A. June-July. Ht. 100cm. **P. *l. Mother's Choice*** has double very large white flowers. Reliable. May-June. Ht. 90cm. **P. *l. Nymphe*** has single, salmon pink flowers. June-July. Ht. 60cm. **P. *l. Karl Rosenfield*** has sumptuous, dark red, semi-double flowers. June-July. Ht. 100cm. **P. *l. Red Charm*** has waxy dark red double flowers in June. Ht. 80cm. **P. l. *Sarah Bernhardt*** has voluptuous, pink, double flowers with a whitish edge to the petals. Ht. 60cm. **P. *l. Scarlet O'Hara*.** Single scarlet flowers. Very early flowering. Ht. 90cm. **P. *l. Shirley Temple*** has large fragrant rose pink flowers which soon turn to white. Wonderful cut flower. June-July. Ht. 100cm. **P. *l. Sorbet*.** Light pink flowers with canary yellow centre. May-June. Ht. 70cm. **P. *Sword Dance*.** Japanese form. Brilliant red petals with dense gold petaloids. May-June. Ht. 80cm. **P. *l. White Wings*.** Huge, single, white flowers. May-June. Ht. 80cm. **P. *l. Yellow Crown*.** Semi double bright yellow flowers with red markings. Dwarf herbaceous/tree paeony hybrid. Sun or part shade. Any well drained soil. Bred by Itoh-Smirnow 1974. **P. *officinalis Rubra Plena*,** is a magnificent old cottage garden plant with heavy headed, deepest crimson, double flowers. May-June. Ht. 70cm. **P. *o. Alba Plena*** has pure white double flowers and is rather taller – 90cm. **P. *tenuifolia*** has deep red flowers and ferny green leaves. Ht. 40cm.
PAPAVER. (Papaveraceae). Papaver comes from the Latin 'pappa' meaning food or milk - a reference to the milky latex. "According to classical mythology the poppy was created by Somnus, the god of sleep, to ease Ceres of her cares and cause her to slumber for in her weariness she was neglecting the corn. After her refreshing sleep the corn revived, and that is why Ceres is usually represented as wearing a garland of corn mingled with poppies. From this legend the ancients derived a comforting belief that poppies were essential to the health of corn". (Alice Coats, 'Flowers And Their Histories'). Poppies are ideal plants for the dry or gravel garden. **P. *atlanticum*** is a long-lived alpine poppy 35cm. in height and 25cm. in spread. It has pastel orange, semi double flowers. Its winter foliage is a cheerful, fresh green. A gentle self-seeder. **P. *heldreichii* syn. spicatum** flowers later than the oriental poppies and is a better behaved, though less showy plant. From a rosette of felted leaves, it sends up a single 60cm. stem, covered in elegant felted, pendant buds open soft apricot. June-Aug. Sp. 25cm. Good in the gravel garden. **P. *orientale*.** The Oriental Poppy was discovered growing in Armenia by Tournefort, who sent seed to George London of London and Wise at the beginning of the 18th Century. Oriental poppies are often banished from the

Geranium sylvaticum *(2)*

tidy modern garden, because of their unruly habits. Having flowered, they tend to crash (like badly behaved guests at a drunken party), crushing your summer flowering perennials like so many Louis Quinze pieces of furniture. There is an answer to this; as soon as you see them topple, dash to the potting shed, unsheathe the shears and cut them to the ground, flowers, stems, foliage. In late summer they will resurrect, like sobered up guests in neatly pressed rosettes of handsome foliage. Oriental poppies are, like most poppies, the easiest of plants in well-drained soil. Tap rooted, they will cope with any amount of drought. They all flower from late May through till the beginning of July and should be allowed a spread of at least 40cm. **P. o. *Beauty of Livermere*** not always true to type. Very large blood red flowers. Ht. 150cm. Lovely with H. matronalis Alba. **P. o. *Bonfire*** has Blood red flowers with a black spot. Ht. 100cm. Sp.45cm. **P. o. *Cedar Hill*** has small pink flowers. Useful compact variety. Ht. 50cm. **P. o. *Cedric Morris*** is one of the very best garden plants, but is inexplicably rare. Wonderful large, smoky pink flowers with a huge black central blotch. Ht. 70cm. **P. o. *Eyecatcher*** is a dwarf variety with fire engine red double flowers. Ht. 40cm. **P o. *Forncett Summer*** has single, salmon pink flowers with ruffled edges. Ht. 75cm. Sp. 40cm. **P. o. *Goliath*** has scarlet orange flowers with a black blotch at the base of each petal. Strong upright stems. Ht. 100cm. Sp. 45cm. **P. o. *Helen Elizabeth*** has single pink flowers. Ht. 70cm. Sp. 40cm. **P. o *Indian Chief*.** Large deep mahogany red flowers. Ht. 75cm. **P. o. *John Paul II*.** Vibrant red flowers. June-July. Ht. 70cm. **P. o. *Karina*.** Pink flowers. Ht. 50cm. **P. o. *Lady Moore*.** Soft pink flowers. Ht. 100cm. **P. o. *Kleine Tanzerin*.** Pink flowers with black spots in centre. Ht. 50cm. **P. o. *Manhattan*** has old fashioned stone red flowers. I am told it sometimes repeat flowers. Ht. 85cm. **P. o. *Mrs Perry*** has pale salmon pink flowers with dark blotch. Ht. 90cm. **P. o. *Patty's Plum*** is the colour of faded purple silk. May-July. Quite wonderful! Ht. 75cm. **P. o. *Peter Pan*.** Double orange flowers. May-July. Very floriferous. Ht. 40cm. **P. o. *Picotee*** has white flowers edged with salmon pink. Ht. 80cm. Sp. 40cm. **P. o. *Pinnacle*.** Pink flower with white centre. Ht. 75cm. **P. o. *Raspberry Queen*.** Raspberry Pink flowers. Ht. 75cm. **P. o. *Royal Wedding*** has white flowers with a black centre. Ht. 80cm. **P. o. *Sultana*.** Deep, luminous, raspberry pink flowers. Ht. 60cm. **P. o. *Tiffany*.** Maroon pink flowers with white edges and black middle. Ht. 85cm. **P. o. *Watermelon*.** Vivid cherry red flowers with black blotches. Ht. 90cm. **P o. *Walking Fire*** has orange red flowers. Free-flowering. Stoloniferous. May-July. Dwarf habit Ht 60cm. **P. *x hybridum flore pleno*, syn. Fireball**, has vivid scarlet orange, double flowers in June-July. Ht. 30cm. Good front of border plant.

PARADISEA. (Asparagaceae). **P. *liliastrum*** is native to Southern Europe. It is similar in looks to an Anthericum but has larger and showier flowers. White flowers. Ht. 40cm. Sp. 30cm.

PAROCHETUS. (Leguminosae). **P. *communis*.** A delightful clover. Electric blue flowers. June-November. Tolerates some shade. Ht. 10cm. Sp. 25cm. Any not too dry well drained soil. Spreads stoloniferously. Unlike P. africanus with which it has been confused, P. communis is stone hardy.

PASSIFLORA. (Passifloraceae). **P. *caerulea*.** The Passion flower. Ivory and blue flowers July-Oct. Ht. 400cm. Sp. 400cm. Any soil. Sun. Best on a south or west wall. Wonderfully uncontrollable. In late summer climbs in my bathroom window to offer me its flowers each morning.

Glaucium flavum *(3)*

PENSTEMONS. (Scrophulariaceae). Penstemons get their name from the Greek for fifth stamen, a reference to the prominent fifth sterile stamen. They are a fashionable plant and rightly so. They provide delicacy, both in form and colour, from summer through to autumn, a season when the garden is stormed by gaudy thugs, Helianthus, Dahlias, Inulas, Chrysanthemums. When suited, Penstemons will flower from June to November and provide a subtle grace, which dances airily through the heavy maturity of autumn. Penstemons should be planted in well-drained, fertile soil in full sun. They are not plants for heavy clay or dry sand, but need good, moisture retentive, well-drained soil. Light soil will need enriching with well-rotted compost, and watering will be necessary during the dry summer months. Heavy clay soils will need lightening with coarse grit and any organic matter which is available. Penstemons should not be pruned hard in autumn as this frequently results in die back and loss of plants. This requires considerable strength of will, as unpruned Penstemons are no beauties in winter. Even if the old stems are sprouting, they should pruned back in April to base, as they will flower better on the new season's growth. Mulching in winter round the base of the Penstemon will protect against frost. Always use a loose mulch. Sticky manure will result in lugubrious rotting.

Blue Penstemons. P. ***glaber*** has distinctive waxy, dense basal foliage. Vigorous and free flowering, its flower are purplish blue. It makes a good, hardy garden plant. Ht. 50cm. Sp. 50cm. **P. *heterophyllus Heavenly Blue*** has greenish blue flowerbuds, which, on opening, become a vivid true blue. It dislikes the crowded border and is happiest in the gravel garden or sheltered scree bed. It has narrow, greyish green foliage and makes a small, sparse, twiggy plant. Ht. 30cm. Sp. 30cm.

Mauve and Purple Penstemons. P. ***Blackbird*** is a most elegant plant. It sends up tall, willowy stems, hung with blackest purple, narrow tubular flowers. It is floriferous and one of the best Penstemons for cutting for the house. Ht. 120cm. Sp. 60cm. P. ***Charles Rudd*** has vivid, lightish purple flowers, which are white throated; it makes a compact plant. Ht. 60cm. Sp. 50cm. Of the large mauve Penstemons I prefer P. ***Lady Alice Hindley***, a magnificent plant, bearing the tall spikes of long tubular, cool mauve flowers with white throats. Not the most accommodating of Penstemons, it does not thrive in poor dry soil and needs rich feeding to perform well. It is alas, also, not reliably hardy and needs protection in winter. But despite these difficulties an indispensable plant. It looks superb with Potentilla recta Sulphurea. Ht. 90cm. Sp. 60cm. P. ***Midnight***. Dark purple flowers. White throat with purple pencil lines. Ht. 70cm. Sp. 50cm. **P. *Raven***. Large, bell shaped, dusky, deep purple flowers. Deep green leaves. Ht. 60cm. Sp. 50cm. P. ***Sour Grapes*** introduced by Marjorie Fish in the 1950s has deep purple flowers with a white throat, striped with purple lines. It makes a compact bushy plant and is a hardy and rewarding border plant. Ht. 60cm. Sp. 50cm.

Pink Penstemons. P. ***Appleblossom*** has small flowers and narrow rather dark green foliage. Not the most vigorous of plants. It needs plenty of light and air. Ht. 60cm. Sp. 40cm. **P. *Evelyn*** is a delightful, small flowered, pastel pink variety, which makes nice compact bushy plants of fresh narrow green foliage. It is easy and reliable. Ht. 60cm. Sp. 50cm. **P. *Hidcote Pink*** has medium sized, shell pink flowers. The upheld throats are elegantly veined with deep pink. A very distinctive robust plant. Ht. 90cm. Sp. 60cm.

Red Penstemons. There are a huge number of named varieties of red Penstemons and awful lot of them are hard to distinguish. Christopher Lloyd recounts how a customer once asked Mr Forbes, a famous Penstemon specialist, for Enlightenment in this matter.

Helianthus Lemon Queen *(2)*

"Tell me Mr Forbes, what is the difference between these two?" "Sixpence" replied Mr Forbes. **P. *Garnet* syn. Andenken an Friederich Hahn** has narrow, green foliage and small, garnet red flowers. Perhaps the hardiest of all the Penstemons; it is also one of the most floriferous, flowering constantly throughout the season, even when not deadheaded. Unlike most cultivars, it succeeds in poor dry soil and preserves the freshness of its foliage to the end of the season. P. Garnet was one of my mother's favourite plants. She planted it with Sisyrinchium striatum. Ht. 50cm. Sp. 60cm. **P. *King George*** bears huge, bell shaped, bright red, white-throated flowers in profusion. A plant of great impact, which lends itself to formal planting. Ht. 50cm. Sp. 60cm. **P. *Windsor Red*.** Abundant, narrow, trumpet shaped, deep red flowers. White throat with magenta pencilling. Narrow leaves. Bushy habit. Ht. 70cm. Sp. 40cm.
White Penstemons. White Hybrid Penstemons are rarities. Perhaps the most beautiful is **P. *hartwegii Albus*.** It has pale creamy yellow buds, which open into long pure white tubular flowers. A very reliable flowerer even in poor summers. Tall growing. Ht. 75cm. Sp. 60cm. Not very hardy but comes true from seed. **P. *White Bedder*** has broader, shorter, whiter flowers than P. hartwegii Albus. Ht. 70cm. Sp. 60cm. Neither P. hartwegii Albus, nor P. White Bedder is very hardy.
PEROVSKIA. (Lamiaceae). **P. *atriplicifolia*** is named after the 19th Century Turkestani statesman B.A. Perovskii. It is a native of gravel screes in Afghanistan and Pakistan and thrives in poor, dry soil, chalky or otherwise. It is salt tolerant and suitable for marine gardens. Perovskia grows to over 120cm. and its silver stems provide strong vertical lines in the late summer herbaceous border. Lavender blue flowers. August-October. Architectural in the gravel garden. Ht. 125cm. Sp. 30cm. **P. *a. Filligran*** has a weak constitution and is not worthy of garden space.
PERSICARIA. (Polygonaceae). The English name is Knotweed. **P. *bistorta superba*** has handsome, dock like foliage and in May-July produces abundant, clear pink pokers. A native of the Caucasus, it resents drought and needs a rich fertile soil in sun or shade, where it will look cool and lush. Lovely with Ranunculus acris Citrinus, which enjoys the same conditions. Ht. 75cm. Sp. 60cm. **P. *microcephala Red Dragon*.** Red leaf with a metalic sheen. Sterile with a non-running root. Strictly clump forming. Insignificant small white flowers. July. Ht. 60cm. Sp. 50cm. Sun or part shade. Any soil. **P. *virginiana Painters Palette* syn. Tovara filiformis** has large, oval leaves, marbled with cream and green and gently suffused with pink. The centre of each leaf is marked with a strong brick red 'V' sign. An elegant, cool looking plant with a nice habit. Negligible flowers in August. "Rats' tails" - says Beth Chatto. Ht. 60cm. Sp. 30cm. Moist soil.
PHLOMIS. (Lamiaceae). The name Phlomis is derived from the Greek for Mullein, presumably because of the woolly stems and leaves of many species. Apart from P. cashmeriana, a rather miffy plant from the Western Himalayas, which requires fertile soil and a sheltered corner, all the other varieties of Phlomis are tough easy Mediterranean plants which thrive in poor dry soil. **P. *cashmeriana*.** Lilac pink flowers July-Aug. Ht. 90cm. Sp. 40cm. **P. *fruticosa***, the mightiest of the family, makes a shrub 80cm. high and 120cm. across. It has silver foliage and rich yellow flowers borne in whorls. June-July. It looks best sprawling onto gravel or paving. Should be gently tidied up after flowering. **P. *russelliana*** is easier to accommodate. An herbaceous, evergreen plant, it makes dense mats of heart-shaped, furry silver green foliage. Excellent ground cover for very dry soil Its yellow flowers are borne in whorls on 90cm. stems. Christopher Lloyd says they look "grubby", but I like them for being somewhat muted.

Helleborus argutifolius *(6)*

Sp. 50cm. Handsome seed heads look good throughout the winter. **P. *taurica* syn. P. pungens.** Mallow pink flowers June-Aug. Ht. 40cm. Sp. 30cm.

PHLOX. (Polemonciaceae). The word Phlox derives from the Greek word for flame. They are members of the Polemonium family, majestic sisters of our common Jacob's Ladder. Their huge heads of scented flowers are indispensable for the summer border and sumptuous as cut flowers. Interestingly the tall erect stems that I and most modern gardeners see as a virtue, (no need for staking – Hurrah!) Alice Coats sees as a defect; she writes that Phlox's "great demerit" is its "stiffness of growth, which can look handsome, but never graceful". Their tall vertical stems continue to look good till February by which time they have deliciously bleached. Border phlox, botanically, belong to two species: Phlox maniculata and Phlox paniculata. Both are native to North America and extremely hardy. Phlox paniculata was first introduced into England by Dr J Sherrard at the beginning of the 18th Century. Phlox love a rich, moist, cool soil. They thrive in loam or well-drained clay. Dry, sandy or thin, chalky soils must be enriched with bulk quantities of well-rotted farmyard manure or other organic matter. Phlox like good light and do not thrive under trees where they are liable to mildew and will produce miserable flowers. But, although liking an open position, they dislike exposure to the full force of the summer sun. They do best in the North border, where their flowers always look elegantly fresh, when the rest of the world is frazzled and irritable. Phlox grown in the open garden are generally free from pests and diseases. The only problems normally encountered are stem eelworm and powdery mildew. The symptoms of eelworm are a narrowing of the upper leaves, which become reduced to a midrib with a narrow, frilly leafblade. Attacks of eelworm are always fatal and any affected plants should be dug up and burnt. Chemicals used by nurserymen to combat eelworm are too dangerous for amateur use. If eelworm does strike, it is essential not to replant phlox in the same piece of ground for three years, as the eelworms can remain dormant in the soil. Luckily eelworm is rare. I have had my present garden for 17 years and, so far, have thankfully not seen a single case. Powdery mildew is much more common, but not nearly so serious. The symptoms are white patches developing on the leaves and the stems. Affected plants should be sprayed every two weeks with a suitable garden fungicide. Plants afflicted with mildew normally flower perfectly satisfactorily, and mild cases of mildew may be safely ignored. If plants are heavily infected, we often cut phlox right down to the ground; if all old growth is removed, the plant will regenerate with healthy foliage. Established clumps of phlox tend to produce an excessive number of flowering stems with correspondingly small flowers. When the young stems are about 30cm. tall, thin out any weak and spindly shoots, allowing the strongest and healthiest to grow on. You will be rewarded with sumptuous blooms. Phlox normally come into bloom in late June and their first flush of flowers may be finished by early August. Cut all the old flowering stems to the ground. Give the plants a good liquid feed of Tomato fertiliser and you will be rewarded with a second flush of flowers in early autumn.

P. *maculata*. Phlox maculata has glossy, deep green foliage and strongly erect stems. It is very mildew resistant and longer flowering than Phlox paniculata. All varieties are fragrant. Flowering period July-Sept. G. S. Thomas claims P. maculata is not as prone to eelworm as P. paniculata. All varieties have a spread of about 60cm. **P. *m. Alpha*** has lilac pink flowers. Ht. 85cm. **P. *m. Natascha*** has lilac flowers with white stripes.

Helleborus orientalis *(2)*

Ht. 85cm. **P. *m. Omega*** has white flowers with lilac centre. Ht. 100cm. **P. *m. Rosalinde*** has purplish pink flowers. Ht. 80cm.
P. *paniculata*. All varieties are fragrant and flower from July to September. They have a spread of about 60cm. **P. *p. Blue Boy*** has mauve blue flowers. Ht. 100cm. **P. *p. Blue Paradise*.** Large violet blue flowers with purple eye. Ht. 70cm. **P. *p. Bright Eyes*.** Scented pale pink flowers with red centre. July-Aug. Strong growing plant. Ht. 90cm. **P. *p. Darwins Joyce*.** A much smarter, variegated form than Norah Leigh. Dark green leaves with a light cream edge. Pink flowers. Ht. 85cm. **P. *p. Europe*.** Fragrant, white flowers with carmine eye. Ht. 75cm. **P. *p. Eventide*.** Lavender blue flowers with a lilac flush. Dark green leaves. Ht. 90cm. **P. *p. Flamingo*.** Flamingo pink flower with a darker eye. Ht. 80cm. **P. *p. Midnight Feelings*.** Flowerless phlox that develops plume-like clusters of reddish black needle like bracts July-Oct. Ht. 70cm. **P. *p. Miss Pepper*.** Pink flowers with dark eye. Long flowering period. July-Oct. Ht. 70cm. **P *p. Mother of Pearl*** has white flowers with a hint of pink. Ht. 75cm. **P. *Orange Perfection*.** Deep orange flowers. A useful short form. Ht. 50cm. **P. *p. Prince of Orange*** has bright orange flowers. Dark green leaves. Ht. 90cm. **P. *p. Prospero*.** Fragrant mauve flowers. Cool and delicious looking. Ht. 90cm. **P. *p. Sandringham*** has pink flowers. **P. *p. Starfire*** has strong red flowers with rich dark foliage. Ht. 90cm. **P. *p. Uspech*** has purple flowers with a white eye. Ht. 80cm. **P. *p. White Admiral*** has very shapely heads of pure white flowers. Ht. 90cm.
PHORMIUM. (Xanthorrhoeaceae). Good structural plants. Their sheathes of dagger like leaves make a good contrast to softer contours. **P. *tenax Jester*.** Grey green, evergreen, sword shaped leaves with a central pink band. Ht. 100cm. Sp. 60cm. **P. *t. Platt's Black*.** Like a beefier Ophiopogon planiscapus Nigrescens. Black, slightly twisted, flat leaves. Best leaf colour in full sun. Moist rich soil. Ht. 45cm. Sp. 35cm. **P. *t. Purpurea*** makes huge clumps of architectural foliage. Ht. 150cm. Sp. 100cm. **P. *t. Yellow Wave*.** Cream, yellow and green striped leaves. Spectacular in winter. Ht. 100cm. Sp. 50cm.
PHYGELIUS. (Scrophulariaceae). All varieties have glossy, deep green leaves and flower June-Sept. Ht. 90cm. Sp. 80cm. They like moist fertile soil and sun. Cut down in spring like a hardy fuchsia. **P. *Devil's Tears*.** Tubular red flowers. **P. *Moonraker*.** Soft yellow, tubular flowers. **P. *Trewidden Pink*.** Faded pink tubular flowers with red lip and yellow throat.
PHYSOSTEGIA. (Lamiaceae). The Obedient Plant, so called, because the flowers are hinged and you can turn them on the stem. They flower July-Aug. Ht. 60cm. Sp. 40cm. **P. *virginiana Bouquet Rose*** has mauvish pink flowers. **P. *v. Red Beauty*** has red flowers **P. *v. Summer Snow*** is pure white. All are vigorous increasers. The less vigorous **P. *v. Variegata*** has pink flowers and silver variegated foliage. All physostegias have good seed heads. Any not too dry soil. Sun.
PIMPINELLA. (Aciaceae). Bright pink flowers. May-July. Ht. 50cm. Any soil. Sun or part shade. Most umbelliferae are architectural beasts. Pimpinella is a dainty delight.
PLATYCODON. (Campanulaceae). **P. *grandiflorus*** Platycodons were introduced into cultivation in 1754 and get their name from the Greek for broad and bell, 'platys' and 'kodon'. The English name Balloon Flower refers to the flower's most striking feature, their flowerbuds - huge inflated balloons. Reginald Farrer says that in Japan Platycodons grow in such abundance that they make a special autumn spectacle; and that in Japanese classical legend and poem the Platycodon is as common as the rose or violet in western culture. Platycodons are beautiful, whether in bud or flower, and are

Helleborus viridis *(3)*

valuable in the summer border for their late flowering. They need a moist well-drained soil and full sun. If starved they will look scruffy and stunted. Unlike most Campanulas, they are adorned with handsome foliage. Their broad, lanceolate foliage is clean and richly glaucous. Platycodons do not seem to arouse much passion among modern gardeners. It was not always thus. Shirley Hibberd, the author of 'Familiar Garden Flowers' was a great enthusiast. In 1898 she wrote "the lover of hardy plants should give no rest to the soles of his feet or the palms of his hands till he has mastered every detail of their cultivation". **P. g. *Fuji Blue*,** single, deep blue variety, **P. g. *Albus*,** pure white single and **P. g. *Hakone's White*,** exquisite semi double. All grow to about 50cm. tall and 20cm. in spread, flowering June-Aug. **P. g. *Sentimental Blue*** is a good dwarf form. Ht. 25cm. Sp. 20cm, it smothers itself in intense blue flowers. June-Aug. Looks good in a pot.

POLEMONIUM. (Polemoniaceae). The name Polemonium, according to Pliny derives from the Greek for war, 'polemos'. Apparently two kings, who each claimed the glory of discovering the medicinal virtues of the herb, took up arms to settle the disputed question. My source for this as for much other curious information is Ann Pratt's classic "Flowering Plants of Great Britain." The medicinal properties incidentally are very slight! As is the cause of most wars! Polemonium caeruleum is native to Britain and although now rare was common in the Late Glacial period. Polemoniums have been cultivated in England since Roman times. Their English name Jacob's Ladder refers to the pinnate leaves, which ascend the stems like rungs on a ladder. "Jacob dreamed and behold a ladder set up on the earth, and the top of it reached to heaven: and behold the the angels of God ascending and descending on it." (Genesis chapter 28 verse 12). Polemoniums are among the most amenable of plants and will thrive in almost any soil, including limestone. They are gentle self-seeders and soften with their ferny foliage our too deliberate plantings. All flower in April-May. Cut down after flowering, they will quickly produce fresh basal foliage and a second flush of flowers. Beth Chatto suggests planting with Ajugas, Millium effusum Aureum and Aquilegias. They will grow in sun or light shade. **P. *caeruleum*** has clear blue flowers April-Oct. Ht. 45cm. Sp. 25cm; ***P. c. Album*** is its very lovely white form. The following hybrids of **P. *reptans*** all make neat compact plants. Ht. 25cm. Sp. 25cm. They flower from April to October. ***P. r. Apricot Delight*.** Apricot pink flowers. Very choice. ***P. r. Blue Pearl*.** Smoky blue flowers. ***P. r. Northern Lights*.** Clear, deep blue flowers. ***P. r. Lambrook Mauve*.** Soft mauve flowers. **P *reptans Pink Dawn*.** Masses of pinkish mauve flowers with slightly bronzed leaves. Pinker flowers than those of Lambrook Mauve. ***P. r. Stairway To Heaven*.** Intense blue flowers. Vigorous habit. Cream edged foliage with pink flushes. Has a brighter appearance than P. c. Brize d'Anjou. Interestingly variegated forms of Polemonium were popular as bedding in the late 19th Century.

POTENTILLA. (Rosaceae). Potentilla is Latin for little powerful one. It was used as a remedy for malaria. The five fingered palmate form of its leaf was considered to have magical properties. Its common name is Cinquefoil. All the Potentillas listed here will grow in any not too dry, well drained soil. **P. *gracilis*** grows about 40cm. tall, Sp. 30cm. and has chrome yellow flowers. One of the first plants to put on its spring raiment; its rich green foliage in early spring is desirable enough; airy sprays of chrome yellow flowers. April-June. **P. *Arc En Ciel*** has double yellow flowers with red centres. June-Aug. Ht. 30cm. Sp. 40cm. **P. *Flambeau*.** Semi-double, cherry red flowers with pale yellow centres. June-Sept. Ht. 45cm. Sp. 40cm. **P. *Gibsons Scarlet*.** Single sharp red

Hemerocallis Chicago Princess *(2)*

flowers. June-Sept. Ht. 30cm. Sp. 25cm. **P. *Monsieur Rouillard*.** Large crimson double flowers. June-Aug. Ht. 40cm. **P. *megalantha*** is an Alpine from Japan. It is a prostrate plant, which makes a neat clump of trifoliate, (so much for cinquefoil!), foliage. The leaves are a soft grey green, against which the large upturned yellow flowers are shown to perfection. The cupped yellow flowers have orange centres. Any not too dry soil, sun or part shade. June-July. Ht. 15cm. Sp. 25cm. **P. *recta Sulphurea*** has pale, creamy yellow flowers. Deep green, cinquefoil leaves lightly clothe its tall sprays, produced in a continuous succession from May to October. It is of easy cultivation, liking plenty of sun and tolerating drought, though it needs some summer moisture to repeat flower. P. recta Sulphurea is one of my favourite plants. Like Ranunculus acris Citrinus, its pale flowers are a perfect foil for soft mauves, such as Viola Maggie Mott. It is also lovely used to cool strong reds or blues. Ht. 70cm. Sp. 25cm. **P. *x tonguei*** makes large clumps of deep green cinquefoil foliage, from which it throws out long, prostrate sprays of small, apricot flowers. A plant which is never out of flower all summer and autumn; it needs moist fertile soil and will grow in sun or part shade. Ht. 20cm. Sp. 50cm.

PULSATILLA. (Ranunculaceae). Pulsatillas are called pasqueflowers, because they are the flower of the feast of the Resurrection. The flowers of the Pasqueflower were commonly used to decorate the Paschal candle and the rich green dye extracted from their leaves and flowers was used to stain eggs as gifts on Easter Sunday. The household accounts of Edward I reveal that four hundred eggs were dyed and gilded for the Easter festivities! Pasque flowers are lovely in all their parts; their flowerbuds richly covered in silky down, the flowers like deep chalices filled with golden stamens, the seed heads stuffed with long luxuriant awns. Pulsatillas are native to our chalk land meadows, but will thrive in most soils, being found in mainland Europe growing in acid sand. Pulsatilla was once a common wildflower in England. John Hill writing in 1756 describes seeing it cover in spring "with its living purple" parts of the Gogmagog hills near Cambridge. Although tolerant of summer drought, Pulsatillas need plentiful moisture in spring. They are best in an open situation, although they will grow in light shade. They are modest plants, never exceeding 30cm. in height and not much more in spread. They flower March-April. **P. *vulgaris***, the species plant has deep purple flowers. **P. *vulgaris Rote Glocke***, a German hybrid, has superb, rich red flowers, against which the golden stamens positively sing. **P. *v. var. Alba*** has pure white flowers - very chaste and very beautiful.

RANUNCULUS. (Ranunculaceae). The botanical name comes from the Latin for frog, 'rana'. One must suppose the Romans considered the Ranunculus froggy, because many of its species, frog like, luxuriate in a moist environment. The English name for the genus, Buttercup, vividly suggests the colour of our native field buttercup, a rich buttery yellow, not the most fashionable of colours, particularly amongst female gardeners, who, it would seem, view strong yellow as some kind of optical obscenity. **R. *aconitifolius Pleniflorus*.** "Fair Maids of France" This exquisite plant, thought to have been brought to England by the Huguenots in the 16th Century, may be froggy in its love for a humid environment, but buttery it is not in its complexion. "The most perfect, pure white, densely double buttons imaginable" (Graham Stuart Thomas). May-July. Dark green, deeply lobed leaves. Ht. 60cm. Sp. 40cm. Moist soil. Sun or part shade. **R. *acris Citrinus*** has flowers of the palest, most delicate sulphur yellow. The petals have a soft sheen and the myriad of flowers borne by a well-established plant shimmer like a glimmering pool. The effect is enhanced by the flowers being borne on almost invisible

Hemerocallis James Marsh *(2)*

30cm. stems, the flowers appearing to be empowered with the magic of levitation. R. acris citrinus flowers June-Aug. Sp. 30cm. Although happiest in moist soil, it performs well in anything except the driest sand. The flowers are enhanced by handsome, highly cut rosettes of foliage, which have bold dark zonal markings. Like all sulphur coloured flowers, it is a brilliant foil for stronger colours. It is superb planted with any of the blue hardy geraniums, for instance Geranium sylvaticum, or the strong pink Geranium endressii. It also looks seductive with mauve flowers, such as Viola Maggie Mott, or Penstemon Alice Hindley. Lovely also with pale pink flowers such as Valeriana officinalis or Persicaria bistorta Superba. **R. *acris flore pleno.*** According to Gerard this plant was a sport discovered growing wild near Latham in Lancashire in 1597 by one Master Thomas Hesketh, a herb collector, who introduced it into London gardens. Rich dark yellow pompon flowers on light airy stalks. May-July. Ht. 90cm. Sp. 20cm. Best in cool moist soil. Like many other plants with small double pompon flowers it has been given the name Bachelor's Buttons. Beth Chatto recommends planting it with Iris sibirica. Unlike some buttercups R. flore pleno is never a nuisance in the garden. It is sterile and never self seeds. **R. *ficaria Brazen Hussy.*** Dark bronze foliage with brassy yellow flowers. Suitably showy for a plant introduced by Christopher Lloyd. March-April. Pops up, struts its stuff and disappears. Sun or part shade. Any not too dry soil. Ht. 5cm. Sp. 30cm.

REHMANNIA. (Orobanchaceae). **R. *angulata.*** Named after Joseph Rehmann, a 19th Century St Petersburg physician, Rehmannia, a native of China, was introduced in about 1890. It is sometimes called The Chinese Foxglove, though apart from the shape of the flowers, this plant has little to suggest the common, native Foxglove, Digitalis purpurea. The rich green leaves are deeply lobed and much more exotic looking than anything you might find in an English hedgerow. The flowers a warm mauve pink are produced on 60cm. spikes in great profusion over a long period. Sp. 40cm. Plants overwintered in a cold greenhouse, where they may be kept almost dry, will come into flower in April and continue in bloom till late November. Rehmannia is a plant of great distinction, exotic in its looks and beautifully proportioned. In the average garden Rehmannia angulata is unlikely to prove itself hardy. It will tolerate any amount of cold, but hates our English winter wet. Sun or part shade. Architectural in pots.

RHODANTHE. (Asteraceae). **R. *African Eyes.*** Lovely miniature white daisy flowers with maroon centre. May-Nov. Hardy and lime tolerant. Finely cut silver foliage. Neat and compact. Sun. Any well drained soil. Ht. 20cm. Sp. 25cm.

ROSMARINUS. (Lamiaceae). The Romans are said to have given the name Rosmarinus (literally "dew of the sea") to this plant because they observed how happily it grew close to the Mediterranaean. It was also the Romans who first introduced Rosemary to Britain. One of the three best plants to grow by a doorway - the others are Lippia citriodora and Myrtus communis - pluck a sprig as you sally forth to work, wonderful to sniff in the odious auto. All Rosemaries flower in June and require well-drained soil and full sun. All are drought tolerant. **R. *officinalis.*** Violet blue flowers. Grey green evergreen, aromatic leaves. Informal habit. Ht. 30cm. Sp. 50cm. **R. *o. Miss Jessop's Upright*** has light blue flowers and a very compact upright habit. Much used for hedging or as a formal statement. **R. *Capri*** has medium blue flowers and a prostrate habit. Ht. 30cm. Sp. 50cm. **R. *Green Ginger*** has ginger scented foliage. Ht. 80cm. Sp. 50cm.

Hemerocallis Wychford *(2)*

RUDBECKIA. (Asteraceae). Rudbeckias were named by Linnaeus after a professor of botany at the university of Upsala, whose greatest academic achievement was his discovery that the Paradise of Scripture was situated somewhere in Sweden!
R. *fulgida Goldsturm.* An easy plant to please. Needs sun and will grow in any soil, which is not completely dry. Clean, dark green foliage. Sturdy 60cm. flowering stems. Sp. 50cm. Reflexed yellow flowers with a handsome black central cone. July-Oct.
R. *maxima.* Graham Stuart Thomas writes: "There are many Rudbeckias with yellow flowers and black centres but there is nothing like this one". Smooth, blue, broad leaves. Yellow reflexed flowers with huge prominent black cones. Aug-Oct. Ht. 150cm. Sp. 50cm. **R. *nitida Herbstsonne.*** I find most cone flowers with their stiff daisy flowers and aggressively upward thrusting black cones too insistent. R. n. Herbstsonne is different; its gently snub nosed, green cones and informally reflexed pure lemon yellow petals are languidly elegant. Aug-Oct. Ht. 175cm. Sp. 60cm. Strong stems. Good cut flower. As Beth Chatto suggests - lovely just by itself, a great jugful.
R. *occidentalis Black Beauty* is definitely one for the flower arranger. Large black cones ringed with minute yellow petals, set off by large green bracts. July-Oct. Ht. 125cm. Sp. 30cm.
RUTA. (Rutaceae). **R. *graveolens Jackmans Blue.*** A Mediterranean herb which was valued by the Greeks as an antidote to poison. Introduced into England by the early 15th Century. Was still valued as an anti-pestilential in the 18th Century. As Alice Coats relates: "the law courts were strewn with it and it was included in the nosegays carried by judges as a protection against jail-fever". Neat mound of intensely blue, finely cut, evergreen foliage. Leaves have for me a strongly erotic aroma. Negligible flowers. Ht. 45cm. Sp. 45cm. Sun or part shade. Any well drained soil. Cut to the base in spring, to ensure plants remain fresh and compact.
SALVIAS. (Lamiaceae). The name is derived from the Latin 'salvare', meaning to heal, Salvia officinalis being at one time credited with medicinal qualities. Evelyn wrote "it is a plant with so many and wonderful properties that the assiduous use of it is said to render men immortal." Salvias are a huge genus and for convenience I have divided them into three groups, biennial salvias, perennial herbaceous salvias and shrubby salvias.
S. *turkestanica* and its monocarpic sister. Salvia turkestanica, sometimes called Vatican sage, is one of the most architectural of plants. It is up there with the best: Acanthus spinosus, Crambe cordifolia, Geranium palmatum, Onopordum acanthium. It is an imposing plant, often growing to as much as 75cm. in height. Sp. 35cm. The flowering stem is much branched and covered with flowers to its base. The flowers' most prominent feature is their large bracts, which are a pinkish mauve. Flowers are produced June-July, but the withered flower heads retain a ghostly beauty through into the autumn. Salvia turkestanica is a biennial and in its first year produces a magnificent rosette of hairy grey green leaves, which is a splendid ornament to the winter garden. Being a biennial should not be a deterrent, this plant being a willing and reliable self-seeder. Salvia turkestanica is an easy plant to cultivate. It thrives in poor, dry sandy soil and is an obvious choice for the gravel garden. To my mind it looks best, planted in an east facing position, which can be viewed from the house. The low early morning sun shining through the pink bracts is a magical sight. Like all salvias, Salvia turkestanica is strongly aromatic. Some people describe it as smelling of tomcat. My brother in law, playing on its English common name, Vatican Sage, has nicknamed it "Pope's Sweat".

Iris Ambroisie *(2)*

Miss Jekyll, Edwardian snob that she was, referred to its aroma as "housemaid's armpit", (downunder that has apparently been transmogrified into "barmaid's armpit"). Similarly, I have heard it described as "hot housemaid". Oh that we might know in what terms it was described by Miss Jekyll's housemaid: "Old Jekyll"? or "Eau de Gerty"? I myself having odd tastes in such matters - for instance I find the smell of Crown Imperial bulbs to be wildly erotic - find nothing obnoxious in its aroma, rather an attractive spiciness. Another salvia closely related to Salvia Turkestanica is ***S. argentea***, which has a wonderful winter, soft silver rosette, covered in a multitude of small hairs. White flowers June-July. Ht. 60cm. Sp. 30cm.

Herbaceous Perennial Salvias are essential providers of blue to the summer and autumn garden. **S. *forsskaohlei*** is indigenous to the Turkish and Balkan coast. It has long spires of blue, white throated flowers, and huge, dark green, heart shaped basal foliage. If cut back to the ground after its first flowers are spent, it will quickly burst back with new foliage and a new flush of flowers. June-Aug. Ht. 90cm. Sp. 50cm. Tolerates drought. June-August. **S. *guarantica Blue Enigma*** bears wonderful, dark, gentian blue flowers on 90cm. spikes. Sp. 50cm. July-Dec. Needs rich, moist well drained soil. Native to South America. Hardy. It is a plant, which, like Perovskia, provides valuable upright lines for the autumn garden. **S. *hians*** comes from Kashmir and was described by William Robinson as one of the best border salvias. It has large, sage coloured, hairy leaves and exotic furry, mauve and white flowers. June-July. It makes a sturdy 60cm. tall plant. Sp. 40cm. **S. *nemerosa Marcus*** is, I am told, creating huge excitement in North America. It bears purple violet flowers from May till Aug. Very compact Ht 25cm. Sp. 40cm. **S. *n. Rose Queen*** has spikes of rose coloured flowers May-Aug. Ht. 60cm. Sp. 25cm. **S. *n. Schwellenberg*** is very different from other S. nemerosas. Both the flowers and the bracts are a vivid wine pink, the overall effect being of much denser colour than in other pink nemerosas. June-Sept. Ht. 40cm. **S. *x superba Blaukonigin*** bears royal blue flowers from July to Sept. Ht. 45cm. Sp. 40cm. All the salvia superbas need moist fertile soil to perform well. **S. x *sylvestris Mainacht*** has spikes of deep purple blue flowers on black stems. May-July. Ht. 60cm. Sp. 30cm. ***S. patens*** has spectacular spikes of gentian blue flowers, which look well with Cosmos atrosanguineus. Ht. 50cm. Sp. 40cm. It flowers July to November. It is interesting to read of John Ruskin's distaste for this plant. He complained of "the lack of moderation in its hue" and that there seemed to be "no gradation or shade" in its "velvety violent blue..... There is no colour that gives me such an idea of violence." He wrote of it being the visual equivalent of a "rough angry scream". It seems fitting, that Ruskin, who, it is known shunned sexual intercourse, should have been revulsed by S. patens, a plant for which a bishop might kick a hole in a stained glass window. S. patens is a native of Mexico, it is hardy in well drained soil. Plants should be marked with canes, because they come up late in spring. **S. *p Cambridge Blue*** has intense, sky blue flowers and is identical in habit to the species. There is also a very nasty dirty white Salvia patens called White Trophy; as miserable a thing as the two blue varieties are choice - never be tempted to buy it. Quite different from other varieties of S. patens is **S. *p. Guanajuato***, which has flowers the same colour as S. patens but twice the size. July-Nov. An imposing sight – grown on good soil it can reach 200cm. Be warned it needs staking virtually from birth. Of all blue flowered plants, **S. *uliginosa*** comes the closest to azure. So often nurserymen describe mauve and purple flowered plants as having flowers of true blue, knowing the word is in itself a magic talisman, which will quadruple the sales

Iris Black Tie Affair *(2)*

of a particular plant. A case in point is Salvia x sylvestris Blauhugel, loudly puffed as a breakthrough in the salvia superba type, having, as its German name would suggest, true blue flowers. Rubbish! Its flowers are no more blue than my varicose veins! Salvia uliginosa is a plant to restore your faith in the meaning of the word blue. It is of an utterly joyful colour, with not the slightest sully to its rejoicing. It starts flowering in August and continues through to November, presenting an uninterrupted spectacle. 140cm. stems erupt in succession, topped by small rocket like heads, packed full of blue dynamite. July-Nov. Sp. 60cm. Salvia uliginosa, like salvia guarantica, is a native of South America and needs rich moist well drained soil, and a protective mulch in winter. It is stoloniferous and gently runs. Plant it and bid adieu to November melancholy! **S. *verticillata Purple Rain*** produces large bold clumps of fresh looking foliage. In summer it bears branching stems of purple bottlebrush flowers; this a much better plant than the species with its anaemic mauve flowers. **S. *verticillata Alba*** a very lovely, pure white form. They are drought hardy and will repeat flower if pruned back hard after first flowering. July-Sept. Ht. 50cm. Sp. 40cm.

Shrubby Salvias S. *blepharophylla* has leaden green leaves and scarlet flowers from July to October. Ht. 30cm. Sp. 40cm. An easy hardy plant which spreads stoloniferously, forming large clumps. Good for the gravel garden. **S. *buchanani*.** Exotic furry magenta flowers. Small dark glossy leaves. July-Nov. Tender. Full sun. A choice compact plant. Ht. 50cm. Sp. 25cm. **S. *Christine Yeo*** is a hybrid between S. microphylla and S. chamaedroides. Very hardy shrub with deep violet flowers from July to November. Ht. 65cm. Sp. 35cm. **S. *Maraschino*.** Large bright cherry red flowers. May-Nov. Ht. 50cm. Sp. 40cm. Hardy. Sun. Well drained soil. **S. *microphylla Grahami*** grows on a west wall of my house. It comes into flower in June and, as late as Christmas, can still be covered with innumerable clear red flowers. Salvia microphylla is often described as tender but has survived in my frost pocket of a garden without serious damage. My specimen in fact grows in a mixture of rubble and sand and is underplanted with Iris unguicularis, which enjoys the same sheltered but Spartan conditions. Salvia microphylla is a most rewarding shrub. Fast growing, in the space of two years, it will reach its full height of 150cm. Sp. 70cm. It needs little pruning; a little gentle shaping during the season is normally sufficient. It resents heavy autumn pruning. The narrow fresh green foliage is well proportioned to the flowers and small sprigs mixed with Jasminum nudiflorum, the yellow winter flowering Jasmine, make delightful cheerful, winter posies. The **S. *greggii*** hybrids are rather smaller. They make dwarf shrubs, never more than 60cm. in height. Sp. 35cm. They flower June-Dec. **S. *g. Raspberry Royal*** has flowers of the richest raspberry. Similar in habit to the greggii hybrids is **S. *x jamensis Laluna***, which has the most elegant, cream flowers. The gregii hybrids and S. x jamensis Laluna are similar to Salvia microphylla Grahami in their cultural requirements; they need drainage, full sun and are best planted against a south or west-facing wall. **S. *involucrata Bethelli*** grows to 90cm. Sp. 40cm. It forms a handsome shrub, well furnished with olive green foliage. The leaves are large and somewhat exotic looking, an effect much enhanced by the large flower heads of shocking cerise flowers. Sept-Nov. Looks wonderfully tasteless with Dahlia Bishop of Llandaff. Like all shrubby salvias its wood is brittle, and therefore a site should be chosen, which is well protected from wind. Unlike the other shrubby salvias, it needs a rich moist soil. Cuttings should be taken in autumn as an insurance against hard winters. **S. *i. Mulberry jam*** is a hybrid with smaller heads of darker flowers, deeper green leathery leaves and less brittle wood.

Iris Jane Philips *(2)*

S. ***leucantha*** is very late flowering, often not opening its flower buds till October. Best grown in a pot and brought into the conservatory where it will flower through the winter. Grey veined foliage. Showy, furry mauve flowers. Christopher Lloyd says "this is the species for which I drool most heavily." Tender. Ht. 60cm. Sp. 50cm.

SAMBUCUS. (Adoxaceae). **S. *Black Lace*.** Deeply cut black foliage and pink-red flowers May-July. Ht. 300cm. Sp. 150cm. Can be coppiced. Sun. Any soil.

SANGUISORBA. (Rosaceae). Sanguisorbas are wonderful structural plants for the herbaceous border. They colour well in autumn and always look well planted with tall grasses. They are happy in any not bone dry soil and will tolerate part shade. **S. *canadensis*.** White bottlebrush flowers May-Aug. Ht. 180cm. Sp. 45cm. **S. *obtusa*.** Utterly tasteless furry, lilac pink, bottle brush flowers. (Just needs the hair curlers and fluffy slippers). Very handsome pinnate foliage. July-September. Ht. 120cm. Sp. 60cm. **S. *officinalis Pink Elephant*.** Slim, rosy red bottlebrush flowers. June-Aug. Ht. 175cm. Sp. 50cm. **S. *o. Tanna*.** Narrow, dark red bottlebrush flowers June-Aug. Ht. 120cm. Sp. 60cm.

SANTOLINA. (Asteraceae). The botanical name derives from the Latin 'Sanctum Linum', or holy flax. Why holy and why flax I do not know. Its English name is Cotton Lavender. Santolina was introduced into England from Southern Europe in the second half of the 16th Century. It quickly became popular as an edging plant for knot gardens. John Parkinson wrote in 1629: "the whole plant is of a strong sweete sent, but not unpleasant, and is in many places planted in Gardens to border knots with, for which it will abide to be cut in what form you think best; for it groweth thicke and bushy, very fit for such workes...". Santolinas are plants of the Mediterranean and thrive on any poor dry soil, alkaline or otherwise. They have aromatic foliage and make pleasant rounded shapes. They look good in gravel. Most varieties have harsh, deep yellow flowers. **S. *pinnata subsp. Edward Bowles*** is a compact form with grey green foliage and tasteful creamy white pompon flowers. May-July. Ht. 25cm. Sp. 25cm. Santolinas must be cut to the ground each spring.

SAPONARIA. (Caryophyllaceae). Saponaria derives from the Latin for soap 'sapo'. Stewed up saponaria leaves make a mild detergent and have been used, since ancient times, by fullers to cleanse wool. Saponaria is cultivated today in Syria for this purpose, and is still widely used by textile restorers in this country in preference to harsher, modern detergents. Saponaria officinalis thrives in any moist fertile soil and flowers over an extended season from June through to September. It is sweetly scented and lovely for picking. Ht. 75cm. Sp. 70cm. An old Dorset name is Bouncing Bett, which may refer to this plant's exuberant habit. Stoloniferous, it spreads rapidly. Lax in habit it benefits from some twiggy support. **S. *officinalis Flore Pleno Alba*** has flowers which are delicate pale blush pink. This is a much better plant than the much more common, dreary S. o. Flore Pleno Rosea which has flowers of a particularly unclean pink. **S. *x lempergii Max Frei*** is new to me. Highly recommended by both Bob Brown and Christopher Lloyd. Pale pink double flowers Aug-Oct. Ht. 45cm. Sp. 30cm. Non invasive. Moist soil and sun. **S. *ocymoides*** is native to the mountains of Southern Europe. A prostrate plant it is good in paving or dry walls. It makes a dense mat of small green leaves, which are covered, in early summer with upturned pink flowers. **S. *o. Snowy Tip*** has darker green foliage and chalky white flowers. Both flower May-June and are about 10cm. high and 35cm. in spread.

SAXIFRAGA. (Saxifragaceae). The name comes from the Latin and means rockbreaking. The leathery leafed varieties are valuable, drought resistant plants for

Iris Supreme Sultan *(2)*

paving, dry walls and the gravel garden. The two following varieties both bear airy sprays of tiny pink flowers in May. **S. x *urbium Aureopunctata*** is a variegated form of London Pride; a vigorous plant, its bold creamy variegation is at its best in winter. Ht. 35cm. Sp. 35cm. The parents of S. x urbium are S. umbrosa and S. hirsuta. **S. *hirsuta*** is a Meditaerranean plant and not native to mainland Britain. Surprisingly it is indigenous to the West of Ireland – an occurrence explained by the legend that S. hirsuta miraculously appeared in order to comfort the nostalgia of the monk Bresal who had spent many months in a Spanish monastery and longed to see the Spanish vegetation again. This theory was also used to explain the presence of Arbutus unedo in Ireland. (Alice Coats, "Flowers and Their Histories"). **S. *x primulaize Salmon*** has much more compact, small, dark green rosettes. A neat plant for troughs. Ht. 10cm. Sp. 25cm.
SCABIOSA. (Dipsacaceae). Scabious comes from the Latin word 'scabies', a parasite that lives in the pubic hair. Scabiosa was thought to be a cure. All scabious need sun and moisture retentive soil. **S. *atropurpurea.*** Clusius received seed of this Mediterranean wildflower from Italy in 1591. Our illustration comes from The New Botanic Garden published in 1812. Traditionally considered a flower of mourning, its French name is "Fleur de Veuve". The Reverend Hilderic Friend writing in 1883 says that a bunch of S. atropurpurea was "considered an appropriate bouquet for those who mourn their dead husbands." How fashions change. S. atropurpurea has shed all its funereal associations and is now a chic flower amongst the gliterati. **S. *a Chile Black*** has maroon black flowers. June-October. Ht. 75cm. Sp. 25cm. Half-hardy it needs well drained soil and a sheltered corner. It is easy to propagate from cuttings. **S. *caucasica*** was first introduced into England by Mr George Loddiges in 1803. He raised it from seed collected on Mt. Caucasus. The Caucasian Scabious is a tough garden plant, which requires no special cosseting. It is a great lime lover. **S. *c. Stafa*** has dark blue flowers from June to September. Ht. 70cm. Sp. 30cm. **S. *c. Alba*** has pure white flowers. June to October. Ht. 70cm. Sp. 30cm. **S. *columbaria Nana*** makes a neat mound of deep green foliage. From late spring till autumn it produces an endless succession of blue flowers. Ht. 25cm. Sp. 25cm. **S. *ochroleuca*** is a bit like a miniature version of Cephalaria gigantea, to which it is closely related. It is an elegant branching plant, which cover itself with creamy yellow flowers. June-Sept. Ht. 80cm. Sp. 25cm.
SCHIZOSTYLIS. (Iridaceae) **syn. Hesperantha. S. *coccinea*** is a native of South Africa. It requires moist fertile soil and full sun. It will also grow in bog or shallow water. Despite their S. African origin, hardiness is not a problem. Clumps need frequent division. Schizostylis is one of the great glories of gardens in October and November (in London I have seen them still in full flower after Christmas). They provide a welcome relief to the over dominance of Compositae such as Chrysanthemums, Dahlias and Helianthus in the late autumn period. They are also an excellent plant for flower arranging, passing my thumb test of a good cut flower: they throw a beautiful shadow on the wall! Always a good test of whether a plant has true distinction. **S. *c. major*** has huge clear red flowers and is an essential plant for the hot border. **S. *c. major Alba*** is a dainty petite plant, looks well with Anaphalis triplinervis. **S. *c. Pink Princess*** has elegant, pale pink flowers. They all have a height of about 60cm. and a spread of 40cm.
SCUTELLARIA. (Lamiaceae). The botanical name comes from the Latin for a small shield, which refers to the small crest on the upper calyx lip. The English name Scullcap flower refers to the shape of the seed. They are easy plants, which thrive in any well-drained soil in sun or part shade. Drought tolerant. **S. *incana*** bears light blue and white

Iris Tide's In *(2)*

snapdragon flowers from June to August. Ht. 60cm. Sp. 40cm. **S. *scordiifolia*** spreads itself by means of small white tubers. It covers itself in July-Aug with masses of intense indigo flowers. Ht 25cm. Sp. 40cm.

SEDUM. (Crassulaceae). The botanical name comes from the Latin 'sedere', to sit, a reference to the way alpine varieties clothe stones and walls. Sedums are good plants for the dry garden. Their succulent leaves act as a buffer against the drought. The flowers of all sedums are loved by butterflies. **S. *African Pearl*.** Vibrant red stems. Black leaves and red flowers. July-Sept. Ht. 50cm.Sp. 30cm. **S. *Bertram Anderson*** has leaves which emerge greyish green in spring, later turning matt black. Deep pink flowers Aug-Oct. Ht. 15cm. Sp. 25cm. **S. *Matrona*** has smoky, purplish green leaves, red stems and pale pink flowers Aug-Sept. Ht. 50cm. Sp. 40cm. Looks wonderful with Potentilla Flambeau. **S. *Purple Emperor*.** I agree with Bob Brown, that this is the best dark leaved sedum. Bushy. Black foliage. Red flowers. July-Sept. Ht. 40cm. Sp. 40cm. Sun. **S. *rupestre Angelina*.** Bright yellow foliage. Yellow flowers. Ht. 15cm. Sp. 60cm. From Christian Krees Sarastro Nursery in Austria. **S. *spathulifolium Cape Blanco*** is grown solely for its foliage, which is a milky grey metallic colour. Its flowers are a tedious yellow. Ht. 10cm. Sp. 25cm. Good in gravel. **S. *spectabile Autumn Joy* syn. Herbstfreude** is as much valued for its architectural foliage as its flowers. Throughout the driest summer, it maintains the wonderful freshness of its bluey grey succulent foliage. Pink flowers, which go dark red with age. Aug-Oct. Ht. 60cm. Sp. 60cm. **S. *spect. Iceberg*** is a short compact form with pure white flowers and apple green leaves. Aug-Oct. Ht. 30cm. Sp 40cm. **S. *telephium Red Cauli*.** A super new sedum. Bright red flowers over purple tinted, blue green foliage. Aug-Oct.Ht. 30cm. Sp. 30cm.

SEMPERVIVUM. (Crassulaceae). Wonderful plants for a trough or the gravel garden. Children delight in their neat, symmetrical rosettes. Will grow in any well drained soil, and, although they require a light position, they do not need full sun. All have a height of about 10cm. and a spread of 25cm. Sempervivums have recently become fashionable for planting as a living roof, a purpose for which they are ideally suited as they need the minimum of soil and nutrients. The best technique to establish individual plants on a roof is to bag their roots with a little compost in the toe of an old pair of tights. **S. *arachnoideum Densum*.** The arachnoides group are distinguished by the attractive fine webbing which covers their rosettes in winter. **S. *Alpha*.** Crimson and green rosettes with hairy tipped leaves. **S. *Bloodtips*** has leaden green leaves with blood red tips. ***S. Mrs Guiseppi*** has fresh green leaves with blood red tips. **S. *Engles*.** Young growth brilliant red darkening to deep wine red. **S. *Othello*** has smoky red rosettes. **S. *Wolcotts*** has smoky grey rosettes.

SIDALCEAS. (Malvaceae). Sidalceas derive their name from the names of two other species of mallow, Sida and Alcea. They are natives of North Western America and hardy easy reliable perennials. In form they resemble miniature versions of the hollyhock; from a neat rosette of deep evergreen foliage, they put up spikes, covered in small rosette-like flowers. **S. *Elsie Heugh*** has very pale pink flowers with delicately fringed petals. Ht. 80cm. Sp. 25cm. **S. *Sussex Beauty*.** Glowing satiny pink flowers. July-Sept. Useful for its height, but may need staking. Graham Stuart Thomas calls it "the loveliest of the clear pinks". Ht. 130cm. Sp. 25cm. **S. *Interlaken*.** Carmine red flowers. June-Aug. Ht. 75cm. Sp. 25cm. **S. *Wine Red*.** Wine red flowers June-Aug. Ht. 80cm. Sp. 25cm.

SILENE. (Caryophyllaceae). Silene is the Greek name for the campion family. **S. *uniflora Rosea*** has palest pink single flowers. June-Aug. Glaucous foliage. Good in

Kniphofia Tawny King *(2)*

dry walls or paving. Ht. 15cm. Sp. 20cm. ***S. Zawadskii*** makes handsome rosettes of deep green shiny foliage, which set off the typical white campion flowers. June-Aug. Ht. 20cm. Sp. 25cm. Lovely in gravel.

SILPHIUM. (Asteraceae). **S. *perfoliatum.*** The Compass Plant – so called because the flowers always face South. North American. Thrives in any moisture retentive soil. Yellow daisy flowers Aug-Sept. Ht. 200cm. Sp. 40cm. A good robust structural plant. Can look good with tall grasses.

SISYRINCHIUM. (Iridaceae). Sisyrinchiums are native to North America. Having told customers for years that Sisyrinchiums must be grown in well drained soil, some of my customers inform me they grow some of the dwarf varieties actually *in* their ponds. All Sisyrinchiums thrive in the gravel garden. **S. *Californian Skies.*** Early flowering. Clear blue flowers. May-July. Ht. 17cm. Sp. 20cm. **S. *E. K. Balls*,** from May to November, thanks to its sterility, never ceases to delight with an endless display of rich blue flowers. Ht. 15cm. Sp. 20cm. **S. *Iceberg*** has fresh green foliage and large, pale, ice blue flowers, which it bears from May till October. Ht. 15cm. Sp. 25cm. **S. *Mrs Spivey*** bears delicate white stars throughout the summer. May-July. Ht. 15cm. Sp. 20cm. **S. *Quaint and Queer*** bears numerous snuff coloured flowers from May to September. Looks wonderful planted with Carex buchanani. Ht. 40cm. Sp. 25cm. Sp. 20cm. **S. *striatum*** has fans of architectural sword shaped leaves. Spires of creamy yellow flowers in June-July, followed by shiny black seed heads. Ht. 65cm. Sp. 50cm. **S. *s. Aunt May*** has leaves, boldly variegated with pale cream. It has the usual, small, pale yellow flowers June-July, followed by shiny black seed heads. It has a reputation for being tender, but is stone hardy with us, given perfect drainage. Ht. 65cm. Sp. 50cm

SOLANUM. (Solanaceae). **S. *jasminoides Album.*** The most wonderful, rampant climber for a West or South wall. In a sheltered spot can still be covered in flower in December. Ht. 300cm. Sp. 200cm.

SOLLYA. (Pittosporaceae) **syn. Billardiera. S. *heterophylla*** is an astonishing shrub. In a conservatory will flower throughout the year. Hardy in sheltered warm gardens. Pure, sky blue flowers. Elegant, narrow green leaves. Ht. 150cm. Sp 40cm.

SPHAERALCEA. (Malvaceae). **S. *monroana,*** like so many members of the mallow family, is gloriously floriferous. If suited it will flower from June till November: a profusion of cupped, coral pink flowers, shown off to advantage by its silvery foliage. It flings out long tendril like branches, which ideally suit it to tub culture, the branches spilling down to the ground make a most handsome mound. It is however an energetic plant and to maintain the vigour of a specimen throughout the season it will need regular potting on, or at least turning out of its tub, tearing off some fibrous root and stuffing back in the tub with some fresh compost. Yes, you can be that unceremonious! As well as in tubs, Sphaeralcea can be grown on a dry bank. Sphaeralcea monroana is half-hardy and needs well drained soil in winter. For those with cold gardens, cuttings should be taken in autumn as an insurance. Delay serious pruning till the spring, as the old growth will give it winter protection. In spring it should be pruned hard, as it flowers better on new growth. **S. *m. Hopleys Lavender*** has delicious pale mauve flowers and **S. *m. Hyde Hall*** has pale pink flowers. All varieties have a height of about 40cm. and a spread of about 120cm.

STACHYS. ((Lamiaceae). **S. *byzantina.*** "Lambs' Ears". Silver grey foliage. Mauve flowers. July-August. Makes a large carpet. Drought tolerant. Good in gravel. Ht. 60cm. Sp. 40cm. **S. *b. Silver Carpet.*** As this variety produces almost no flowers, its

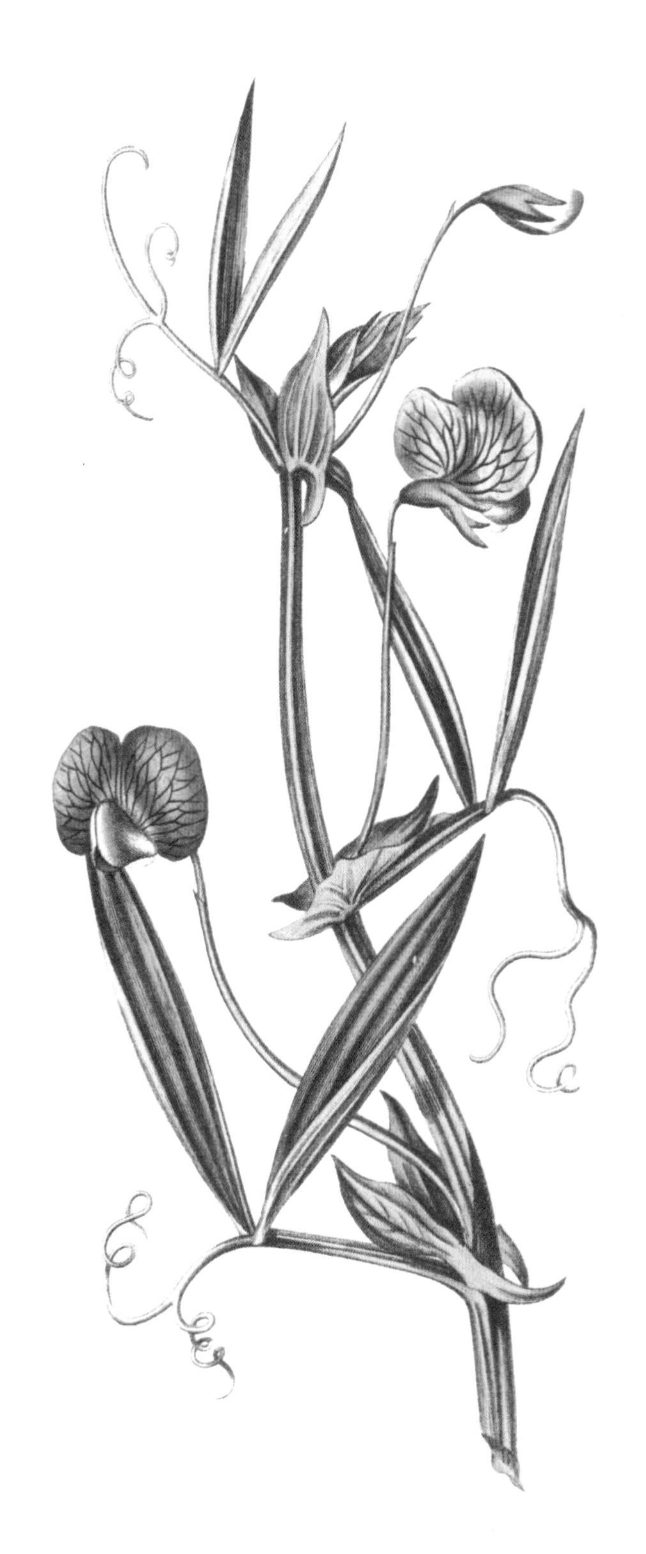

Lathyrus sativus *(3)*

silver lambs ears foliage looks good all summer. Good ground covers for hot dry areas. Ht. 20cm. Sp. 50cm. **S. *grandiflora*.** Whorls of hooded purple flowers on upright stems. July-Aug. Large, rich green, basal leaves. Good ground cover for moist fertile soil. Sun. Ht. 60cm. Sp. 40cm.

STOKESIA. (Asteraceae). Native to the South-East of the United States, Stokesias like sun and moisture retentive soil. **S. *laevis*.** Handsome, deep green rosettes of leaves. Large, cornflower like blooms on 35cm. stems in July-September. Sp. 30cm. **S. *l. Alba*** has cool white flowers. **S. *l. Blue Star*** has deep blue flowers. **S. *l. Mary Gregory*.** A colour form discovered in the wild by Mary Gregory. Large, pale, butter yellow flowers on short stems. **S. *l. Purple Parasols*** has large, deep violet purple aster flowers. **S. *l. Traumerei*** has white flowers with a touch of pink.

STROBILANTHES. (Acanthaceae). **S. *attenuata* syn. S. atropurpurea.** Violet blue flowers. Aug-Oct. Sun. Moist fertile soil. Does not like poor dry sandy soil. Ht. 120cm. Sp. 80cm. Always heavily molested by bees here. I rate this as one of the top ten herbaceous plants. Christopher Lloyd dismisses it as not worth a space in your garden.

SUCCISA. (Dipsacaceae). **S. *pratensis* syn. Scabiosa succisa.** The Devil's Bit Scabious. This vernacular name seems to have originated in a Byzantine legend, which suggests this plant's blunt ended root was bitten off by the Devil. Rich green, basal foliage. Abundant, steel blue flowers on strong 55cm. stems. Late summer. Any soil. Sun. An easy, airy, graceful plant. Sp. 30cm. Gently self-seeds.

TANACETUM. (Asteraceae). The species plant **T. *coccinea*** was introduced from the Caucasus at the beginning of the 19th Century and was soon the subject of intensive plant breeding. At the end of the 19th Century Robinson recorded some fifty varieties. I think Tanacetums, or Pyrethrums as they used to be called, are unjustly neglected. For me they far outstrip in loveliness most of their daisy cousins. Leucanthemums, Chrysanthemums and Asters cannot compare for daintiness of flowers or freshness of foliage. **T. *c. Beauty of Stapleford*** has mid pink flowers. May-June Ht. 70cm. Sp. 40cm. **T. *c. Brenda*** has cerise flowers June-July. Ht. 75cm. Sp. 40cm. **T. *c. Eileen May Robinson*** has clear pink flowers. May-June. Ht. 70cm. Sp. 40cm. **T. *c. Evenglow*** has bright salmon flowers May-June. Ht. 60cm. Sp. 40cm. **T. *c. H. M. Pike*** has bright red flowers. May-June. Strong grower. Ht. 75cm. Sp. 40cm. **T. *c. James Kelway*** has scarlet flowers. May-June. Ht. 75cm. Sp. 40cm. **T. *c. Robinson's Pink*** has deep pink flowers. May-June. Ht. 70cm. Sp. 40cm. **T. *c. Robinsons Red*** has clear deep red flowers. May-June. Ht. 70cm. Sp. 40cm. **T. *c. Vanessa*** has flowers shading from lilac to cerise. May-June. Ht. 65cm. Sp. 40cm. **T. *vulgare* Golden *Feather*.** Very welcome in early spring for its bright gold, much divided foliage. If cut back in summer, produces a second flush of bright new growth. Boring small white daisy flowers. Sun and good drainage. Ht. 35cm. Sp. 30cm. Any not too dry soil.

TEUCRIUM. (Lamiaceae). **T. *hircanicum*.** Introduced from Persia in 1763. Bushy plant of aromatic sage like foliage. Slender spikes of grey buds open to dark lilac flowers. July-October. Ht. 60cm. Sp. 90cm. Good in gravel garden.

THALICTRUM. (Ranunculaceae). All Thalictrums do best in a rich moist cool soil. Happy in full sun or part shade. Lovely in a North border. Valuable as much for their elegant narrow vertical lines and delicate, lacy Aquilegia like foliage, as for the beauty of their flowers. **T. *aquilegifolium*** is native to Europe and temperate Asia but not to Britain. It was definitely in cultivation in English gardens by 1730 as it is mentioned in Tournefort's 'Herbal'. **T. *a. Album*.** Frothy heads of ivory flowers with pink stamens.

Lavatera maritima *(2)*

June-July. Ht. 125cm. Sp. 30cm. **T. *a. Thundercloud.*** Double pinkish blue flowers. June-July. A better doer I am told than Hewitts Double. Ht. 80cm. Sp. 30cm. T. delavayi was discovered by the Abbe Delavayi in Western China in the late 19th Century **T. *delavayi Hewitts Double.*** Heads of light airy purple flowers. Aug-Sept. Ht. 90cm. Sp. 25cm. **T. *flavum Illuminator.*** A selected form of our native ditchlover. The roots of this plant were used to dye wool yellow. Heads of creamy yellow flowers. June-July. Grown mainly for its soft, creamy grey, winter rosettes, which have pink and gold highlights. Unlike most Thalictrums which thrive in part shade T. f. Illuminator needs an open position; planted in shade its flowering stems tend to grovel. Ht. 50cm. Sp. 20cm. **T. *rochebrunianum.*** Heads of light airy purple flowers. Aug-Sept. Ht. 90cm. Sp. 25cm. Moist fertile soil. Part shade. **T. *kiusianum.*** Japanese origin. Produces a mass of small purple frothy flowers on low clumps of finely cut foliage. June-Sept. Ht. 15cm. Sp. 25cm. Sun or part shade. Moisture retentive soil. A useful plant for the front of the woodland border.

THERMOPSIS. (Leguminosae). **T. *lanceolata* syn. T. lupinoides.** Has straw yellow, lupin like flowers May-June. Lovely for fresh green foliage and almost black stems in March. Unlike T. montana, this variety does not run. Strictly clump forming. Tolerates drought. Ht. 60cm. Sp. 25cm.

THYMUS. (Lamiaceae). **T. *citriodorus Silver Posy*** has pretty, grey green and silver variegated leaves. Pale pink flowers in June. Lemon scented foliage. Ht. 30cm. Sp. 30cm. **T. *Doone Valley.*** Yellow and green variegated leaves. Pink flower which are particularly atractive to bees and butterflies. June-July. Ht. 15cm. Sp. 25cm. **T. *serpyllum Russettings.*** Vigorous creeping thyme. Forms a mat of dark green leaves with abundant dark pink flowers all summer. Ht. 10cm. Sp. 20cm. **T. *Hartinton Silver.*** A prostrate thyme with variegated cream and light green foliage. Pink flowers. June-July. Ht. 8cm. Sp. 20cm.

TRADESCANTIA. (Commelinaceae). Introduced from Virginia by John Tradescant Senior at the beginning of the 17th Century. Tradescantias are tough, easy plants. They will grow in any soil in sun or part shade. They are good ground cover and in moister soils will produce an endless succession of flowers throughout the summer, from June to September. Cut back hard after each flush of flowers. Slugs can be a problem. **T. *brevicaulis*** is a handsome thing with cyclamen pink flowers. Ht. 45cm. Sp. 40cm. **T. *Concord Grape*** has bluish grey leaves and violet purple flowers. Ht. 55cm. Sp. 40cm. The flowers and foliage are in perfect harmony. **T. *Innocence*** has pure white flowers. Ht. 60cm. Sp. 40cm. **T. *Osprey*** has white flowers with a blue centre. Ht. 50cm. Sp. 40cm. **T. *Purple Sabre.*** Large broad purple leaves. Pink flowers. Ht. 50cm. Sp. 40cm. **T. *Sweet Kate* syn. Blue And Gold.** Purple flowers with yellow foliage. Nice for brightening up a dark corner. June-Sept. Ht. 60cm. Sp. 30cm. **T. *White Doll*** is a very good new dwarf white form, which grows to a mere 25cm.

TRIFOLIUM. (Leguminosae). The botanical name for clover comes from the Latin for three leafed. **T. *Harlequin.*** Green leaves with broad silver edge and red markings. Very crisp looking. Ht. 15cm. Sp. 25cm. **T. *repens Purpurascens Quadrifolium*** is rather an imposing name for a modest plant. It is a small creeping thing with dark chocolate foliage. Like all clovers, it does not like to be too dry. Ht. 10cm. Sp. 30cm. **T. *Susan Smith.*** Emerald green leaves finely netted with gold. Pink flowers July-Aug. Cool soil. Good in clay. Ht. 10cm. Sp. 35cm. **T. *Wheatfen.*** Maroon flowers over deep maroon leaves. June-July Ht. 15cm. Sp. 25cm.**T. *William.*** Maroon flowers. June-July.

Lilium candidum *(3)*

Maroon shaded leaves. Good in pots. Bob Brown recommends planting it as ground cover between Eucomis Sparkling Burgundy. Ht. 25cm. Sp. 35cm.
TRITELEIA. (Asparagaceae). **T. *Starlight*.** S. African bulb. For me a plant of supreme elegance. Umbels of pale straw coloured flowers. May-July. Long lasting cut flower. Ht. 30cm. Sp. 15cm. Sun and sharp drainage.
TROLLIUS. (Ranunculaceae). The name Trollius is derived from the Latin for basin "trullius" – a reference to the rounded shape of the flowers. Trollius likes rich moist soil. **T. *chinensis* syn. T. ledebourii** bears magnificent, tattered orange flowers July-August Ht. 90cm. Sp. 40cm. Looks handsome with Campanula Kent Belle. **T. *europaeus*** is a British native flower, which in the wild shuns the southern counties of England. Introduced into London gardens in the late 16th Century. It has sulphur yellow flowers. June-July. Ht. 90cm. Sp. 40cm.
TULBAGHIA. (Amaryllidaceae). Tulbaghias were named after an 18th Century governor of the Cape of Good Hope. Members of the Allium family they have a splendid foxy wild garlic smell. Despite their South African origin, they are stone hardy in well-drained soil with the exception of the half hardy T. violacea Tricolor. They flower without respite from midsummer through to the hard frosts. **T. *violacea*** has mauve flowers. Ht. 35cm. Sp. 30cm. **T. *v. Alba*** - perhaps not quite as vigorous as the mauve; Ht. 30cm. Sp. 30cm., **T. *v. Tricolor***, which is a refined plant with silver, pink and green variegated foliage, lovely in a pot on the terrace in summer, it has the usual mauve flowers. Ht. 30cm. Sp. 25cm. **T. v. *White Star***, a name of our own invention for this lovely plant, which has mauve flowers with a distinct white throat. (Does anybody have the correct name?). Ht. 35cm. Sp. 30cm.
TWEEDIA. (Asclepiadaceae). Tweedia, named after a certain J. Tweedie from Aberdeenshire, is a native of Uruguay. **T. *caerulea*** has flowers of a unique ice blue and blooms from August to November. Ht. 50cm. Sp. 20cm. Needs a warm sheltered position and well-drained fertile soil. Not reliably hardy, but easy from seed.
VALERIANA. (Valerianaceae). **V. *officinalis*.** Valerian possibly derives its name from the Latin 'valere', to be healthy - a reference to its medicinal properties. Valerian tea is taken to calm the nerves. The wild valerian is one of the prettiest of our native plants. It has soft pink flowers, borne on narrow upright 150cm. stems, lightly clothed with apple green foliage. June-July Sp. 60cm. Very fresh looking. As Gerard tell us "in the wild it is found in moist places hard to river sides, ditches and watery pits". In the garden, however it will tolerate a wide range of conditions - everything except the driest soil. Looks lovely with Ranunculus acris Citrinus. **V. *phu Aurea*** is grown for its bright gold foliage, which cheers up the grey days of March. The foliage will gradually fade to green, but if cut to the ground quickly returns bright gold. Worthless flowers in August should be strangled in the womb. Ht. 20cm. Sp. 35cm.
VERBASCUM. (Scrophulariaceae). "Called of the Latines candela regia and Candelaria, because the elder age used the stalks dipped in suet to burne, whether at funerals or otherwise". (Parkinson. 'Theatrum Botanicum' 1640). The Verbascum family is valuable in the garden for its tall vertical lines. They all thrive in poor dry soil and full sun. Relatively short lived, they are reliable self-seeders and most varieties can be propagated by root cuttings. They all have a spread of roughly 30cm. and flower from June to August. **V. *Annie May*** has darkest damson flowers. Foliage and stems also suffused with purple. Reliably perennial. Ht. 90cm. **V. *Apricot Sunset*** has apricot and pink flowers. Reliably perennial. Ht. 100cm. **V. *chaixii album*** has white flowers with

Linaria purpurea Canon Went *(3)*

a plum centre. Ht. 75cm. Deep green foliage. A robust and reliable plant. **V. *Norfolk Dawn*** has large biscuit coloured flowers. Reliably perennial Ht. 80cm. Sp. **V. *Patricia.*** Biscuit, pink and buff coloured flowers open from bronze buds. Ht. 90cm. **V. *Southern Charm*** has coppery pink flowers. Ht. 60cm. **V. *Gainsborough*** has soft yellow flowers and sage green foliage. Very desirable. Ht. 120cm. **V. *Cotswold Beauty*** has coppery pink flowers like a lighter version of those of the much hyped V. Helen Johnson. Unlike V. Helen Johnson, who may be pretty as sin but whose dalliance is definitely ephemeral, V. Cotswold Beauty is a reliable perennial plant. Ht. 120cm.**V. *Violetta*.** Deep violet flowers. Purple flushed green leaves. Ht. 75cm.

VERBENA. (Verbenaceae). All verbenas need sun and good drainage. **V. *bonariensis*** was first introduced to England from Buenos Aires by Dr James Sherrard in 1726, but it took nearly two hundred and seventy years for it to become a popular garden plant! V. bonariensis in its native habitat grows in marshy ground In this country, although summer moisture is relished, it is wise to plant it in well-drained soil. This greatly increases the chances of it successfully overwintering. Although not stone hardy, V. bonariensis will withstand winter temperatures of down to -10°C. If killed by a hard winter, this, like so many of my favourite plants, will self-seed. Do not despair if seedlings emerge very late, often not until the end of May; despite this late start, seedlings will flower the same year. V. bonariensis grows up to 150cm. in height. Sp. 60cm. Its flower stems are much branched and, although in appearance underengineered, they never need support. The flowers are lavender blue. Dwarf Verbenas such as V. Sissinghurst have long been essential plants for adding colour to the summer and autumn border. Until recently most of these have been tender. The following new dwarf hybrids are all rock hardy in our well drained soil. They are all about 35cm. tall and make large spreading plants – perhaps 60cm. in diameter. They flower from July to November. They prefer sun but flower surprisingly well in part shade. **V. *Edith Edelmann*.** Endless bright clear red flowers **V. *Hammerstein Pink*.** Clear shell pink flowers. **V. *La France*.** Endless large luminous mauve flowers.

VERONICA. (Plantaginaceae). The English name is Speedwell and in Ireland the plant was sewn onto a traveller's clothes to keep him from harm. The wild speedwell Veronica chamaedrys is a plant of the roadside. As a family Veronicas have grabbed some of the best blues. **V. *austriaca*** is as its name suggests a native of Germany and Austria. Gerard grew it in his garden in the late 16th Century. All varieties of V. austriaca are vigorous plants and grow about 25cm. tall with a spread of about 35cm. V. austriaca requires an open position and moist fertile soil. **V. *a. Crater Lake Blue*.** Intense, violet blue flowers. ***V. a Knallblau*** has mid blue flowers. ***V. a. Shirley Blue*.** Darkish blue flowers. **V. *a. Royal Blue*** bears bright mid blue flowers Ht 25cm. Sp 35cm. **V. *gentianoides*** is a native of the Caucasus and was introduced in 1784. **V. *gentianoides Robusta*** produces its first, lovely, scrubbed blue flowers in May. It has rich green, glossy, evergreen foliage and looks good with Geums, with whom it shares a partiality for moist fertile soil. May-June. Ht. 50cm. Sp. 40cm. **V. *g. Robusta*** is an improved form, which has a much longer flowering season than the species, flowering well into August. **V. *g. Tissington White*** is an elegant white flowered variety which has particularly smart foliage and a compact habit. **V. *g Variegata*** has white and green variegated leaves. **V. *peduncularis Georgia Blue*** is a lovely introduction by Roy Lancaster from the Caucasus. This very special speedwell has intense blue flowers and small, evergreen bronzed leaves. Its habit is semi prostrate and it will happily colonise a sunny bank, or fall over the edge of a wall. Its

Lobelia cardinalis *(3)*

flowering period is a matter of some curiosity. When it was first introduced, Roy Lancaster wrote it up in the R. H. S. journal as summer flowering. We grew it for ten years and it never flowered in summer. But much more welcome, it always flowered from autumn through the winter into spring. Only the very cruellest frost can temporarily dampen its spirits, but as soon as the weather relents, this plucky little Veronica will recommence flowering. It has always been in flower here on Christmas day. Veronica peduncularis should be planted in well-drained soil. It dislikes frost pockets. It enjoys baking sun and looks splendid on the rockery, which in early winter looks so naked. In spring, after it finishes flowering, it should be given a brutish haircut and forgotten about till the autumn. Oct-April. Ht. 15cm. Sp. 40cm. **V. *petraea Madame Mercier*.** Glossy green leaves. Blue flowers June-Sept. Ht. 15cm. Sp. 25cm. **V. *prostrata*.** Violet blue flowers May-Aug. Ht. 10cm. Sp. 25cm. **V. *spicata*** is a rare native of Britain. **V. *spicata Erika*.** Pink flowers on striking red stems. Ht. 50cm. Sp. 30cm. **V. *Royal Candles*.** Deep violet blue flowers July-Aug. Foliage remains looking good even when in flower – unusual for a V. spicata. Ht. 45cm. Sp. 30cm. **V. *Trehane*.** Chartreuse foliage. Deep blue flowers. Ht. 25cm. Sp. 35cm.

VERONICASTRUM. (Plantaginaceae). Veronicastrums are valuable for their erect lines. **V. *virginicum Album*** has pure white flowers July-Sept. Ht. 120cm. Sp. 30cm. **V. *Pink Glow*.** Clear pink flowers Aug-Sept. Ht. 100cm. **V. *v. Lavendulturm*** has lavender blue flowers August-September. Ht. 150cm. Sp. 30cm. Any soil, which is not too dry. Sun. Piet Oudolf plants Veronicastrums with tall grasses. **V. *villosulum*.** A good one to test your learned friends on. I bet none will recognise this as a Veronicastrum. Long trailing stems up to 180cm. long with small clusters of deep purple flowers. June-Oct. Good for trailing over walls.

VIOLA. (Violaceae). **Viola cornutas** are invaluable plants for the edge of a sunny border. They will grow happily in any but the most arid of sandy soil or the heaviest clay and are soundly perennial. They spread by stolons and quickly make a good clump. The only attention they require is an occasional chopping back of tired growth in order to make way for the new basal growth. They look good in window boxes or in small pots or urns. They are all about 15cm. in height and make plants about 25cm. in spread. They flower from April through to August. **V. *cornuta Alba*.** Small white flowers in profusion. **V. *c. Boughton Blue*.** Soft lavender blue flowers. **V. *c. Netta Strachan*.** An unceasing succession of pale mauve flowers. **Viola Hybrids.** These are similar in habit and culture to the cornutas but not all have inherited the cornutas sound perennial habit. **V. *Dawn*.** Violetta. Primrose yellow flowers. Floriferous. Good perennial habit. Raised by Howard Crane. **V. *Etain*.** Creamy yellow flowers with lavender margin. Highly scented. Good perennial habit **V. *Irish Molly*.** Bronzed gold and maroon flowers on a yellow base. Needs renewing regularly from cuttings. **V. *Maggie Mott*** has large soft silvery mauve flowers with the palest cream edge. Highly fragrant. Greatly loved by my mother, who remembered it growing under the appletrees in the orchard at her Grandfather's house in the 1920s. Not reliably perennial. Needs renewing from cuttings each autumn. **V. *Milkmaid*** has small creamy white flowers flushed with the palest mauve. Good perennial habit. **V. *Molly Sanderson*** has medium sized jet black flowers with a yellow eye. Reasonably long lived but best renewed from cuttings every couple of years.

WATSONIA. (Iridaceae). S. African in origin, **Watsonia** like moist rich soil, sun, and a frost-free position. Dramatic gladioli like flowers in early summer. Pinks, reds, and maybe an odd white. Ht. 60cm. Sp. 35cm.

Lychnis chalcedonica *(3)*

ZANTEDESCHIA. (Araceae). Zantedeschias are named after the 19th Century botanist, Giovanni Zantedeschii. They are native to South Africa. My father used to call them pig lilies, because, in Ireland, arum lilies and pigs are to be found nuzzling up to each other in muddy ditches. He did not like them. I think he thought, by claiming the name of lily they blasphemed against true lilies, such as Lilium candidum, which for him was a sacred thing, in all its trumpet glory, broadcasting the Annunciation eternally. For me the Zantedeschias have their voluptuous appeal, the absolute white nakedness of their spathes against the lush deep green leaves. They all need moist conditions and an open situation to do well. They are classic plants for the bog garden and can be planted in baskets in ponds. They flower in July-August. **Z. *aethiopica Crowborough*** has large white flowers and grows to 90cm. in height. Sp. 50cm. **Z. *Green Goddess*** has strange, mottled green spathes with white throats and magnificent foliage. Ht. 150cm. Sp. 50cm. **Z. *Kiwi Blush*.** At last a hardy arum which has learnt to blush. I prefer my arums in unapologetic white nakedness, but for those who find modesty charming - pink flowers July-August. Ht. 90cm. Sp. 50cm.

Mertensia virginica *(3)*

Hardy Geramiums

GERANIUM. (Geraniaceae). Hardy Geraniums gained both their botanical name and vernacular English common name from the likeness of their seeds to a Crane's bill. Cranesbills are rewarding and easy garden plants. Unless otherwise stated, in the individual notes below, they will grow in sun or part shade, and will tolerate any soil with the exception of bog.

Hardy Geraniums for Sun

G. *Ann Folkard*. Young leaves lemon yellow. Magenta flowers with a dark eye. June-Sept. Ht. 30cm. Sp. 100cm. **G. *Bob's Blunder*** has pewtered bronze leaves and big soft pink and white flowers. May-Nov. A robust plant. Easy even in heavy clay. Ht. 30cm. Sp. 50cm. **G. *Blue Cloud*** bears pale blue flowers veined with crimson. May-Sept. Finely divided leaves. Ht. 65cm. Sp. 40cm. Sun. Well drained soil. **G. *cinereum Guiseppii*.** Veined mauve flowers. May-July. Ht. 15cm. Sp. 30cm. Full sun. Well-drained soil. The most vigorous of the cinereum group. **G. *c. Lawrence Flatman*** is the daintiest of hardy Geraniums. Greyish green cut foliage. Prostrate habit. Deep lilac flowers flecked with crimson. June-Sept. Ht. 10cm. Sp. 20cm. Dislikes clay. Good in gravel or paving. **G. *c. Purple Pillow*.** Rich purple flowers with black stripes. April-Sept. Blue green foliage. Ht. 15cm. Sp. 25cm. **G. *clarkei Kashmir White*.** White flowers with lilac veins. **G. *c. Kashmir Pink*.** Dusky pink flowers. **G. *c. Kashmir Blue*.** Clear mid blue flowers. All varieties of G. clarkei flower July-Aug and have a strongly spreading stoloniferous habit. Ht. 60cm. Sp. 40cm. **G. *Coombland White*** has marbled leaves and pale pink, red veined flowers, which turn white with age. June-Sept. Ht. 20cm. Sp. 30cm. **G. *Dusky Crug*** has broad chocolaty grey foliage with large sugar pink flowers May-Dec. Ht. 20cm. Sp. 30cm. Happy on any soil including clay. Bob Brown claimed for this Bleddyn Wynne-Jones introduction the title of "best new hardy geranium hybrid". I agree. It has the longest flowering period of any geranium. Looks particularly stunning with Nerine bowdenii in October-November. **G. *himalayense Gravetye*** has deep blue flowers, veined with reddish centres. Moderately lobed leaves. The most free flowering of all the blue himalayense Cranesbills. May-Oct. Spreads by rhizomes. Ht. 30cm. Sp. 50cm. **G. *himalayense Johnson's Blue*** is a sterile hybrid between G. pratense and G. himalayense. Flowers are perhaps the purest blue of all Cranesbills. May-July. Ht. 35cm. Sp. 50cm. **G. *Joy*** (traversii var. elegans x lambertii). Large, glistening, pink flowers with red veins and centre. June-September. Elegant marbled foliage. Ht. 40cm. Sp. 30cm. **G. *maculatum Album*** has the cleanest chalk-white flowers displayed against deep green foliage. May-June. Ht. 50cm. Sp. 40cm. **G. *maculatum Expresso*** has dark brown foliage and bright pink flowers June-July. Ht. 35cm. Sp. 40cm. **G. *maculatum Chatto's Pink*** has clear, shell pink flowers. May-June. Ht. 50cm. Sp. 40cm. Both forms of maculatum are good for naturalising in woodland or field and

Mimulus cardinalis *(1)*

colour well in autumn. **G. *maderense*.** Like G. palmatum but finer, more deeply cut leaves, red stems and bigger in all its parts. A mature plant forms a small trunk. Magenta flowers with a dark eye. May. Ht. 120cm. Sp. 100cm. Monocarpic. Tender. Conservatory in most parts of England. Hardy on South Coast. **G. *x magnificum*.** A vigorous hardy Cranesbill. Large deep blue flowers in June. Attractive hairy foliage. Good autumn colour. Sun or part shade. Hybrid between G. platypetalum and G. ibericum. Sterile. Looks magnificent with orange oriental poppies. Ht. 50cm. Sp. 60cm. **G. *nervosum Nimbus*** has deep violet flowers. May-Oct. Deeply divided leaves, which emerge gold, later turning green. Ht. 40cm. Sp. 50cm. **G. *orientalitibeticum*** is a charming, dwarf plant, which spreads by small tubers. It has green leaves with decorative, darker mottling and large, upturned, white throated, deep pink flowers in May-July. Ht. 20cm. Sp. 25cm. **G. *oxonianum Lace Time*** is one of the shorter growing varieties in its group. It has white flowers, with handsome pink veins May-Aug. Ht. 45cm. Sp. 45cm. **G. *palmatum*** was introduced in the late 18th Century. Norwich Castle Museum has a very handsome, china Derby plate, depicting the plant under its old name Geranium anemonefolium. Geranium palmatum is one of the six great, architectural, hardy herbaceous plants! It is something of a mystery that, despite its ease of cultivation, it is still so little known in English gardens. The root of the problem may lie in that for many years it was much confused with G. maderense, which, of course, is tender and monocarpic, qualities, which were falsely ascribed to G. palmatum. G. palmatum is in fact extremely hardy in well-drained soil and, although not long lived, certainly cannot be termed monocarpic. Its lack of longevity moreover is not a problem as it reliably self-seeds. Although the tall sprays of mauve flowers are both most decorative in their individual flowers and elegant in their overall construction, the true glory of this plant is its foliage. G. palmatum throws out from a central crown large, frond-like, deeply divided leaves, which form a majestic, evergreen mound. It is an easy plant to grow and will accommodate itself to most soils, which do not become waterlogged in winter. It is possibly most effective in part shade, where its flowers will last for months and its foliage will be of a richer and more lustrous green. June-Sept. Ht. 100cm. Sp. 60cm. **G. *Patricia*.** G. endressii x G. psilostemon hybrid. Large magenta flowers over a long period. Differs from G. psilostemon in "having had the sting taken out of the magenta" (Christopher Lloyd comments disaprovingly) and in a much extended flowering season. May-Oct. Ht. 75cm. Sp. 60cm. **G. *Philippe Vapelle*.** Hybrid from G. renardii. Flat blue flowers. June-July. Perfect mounds of sage green leaves. Ht. 35cm. Sp. 30cm. **G. *pratense*.** Pratense means meadow and this plant is indeed known in English as the Meadow Cranesbill, although, truth be told, in the wild it is as like to be found in open woodland as in pastures. Geranium pratense was cultivated by Gerard in his famous garden at Holborn in the late 17th Century. Unlike most Cranesbills, it has a decided preference for moist, fertile soil. All varieties have handsome deeply divided leaves. Mildew can sometimes be a problem with Geranium pratense towards the end of its flowering period. The remedy is simple. If ruthlessly cut to the ground the plant will instantly produce a new rosette of uninfected foliage. The single forms flower in May-June. If cut to the ground after flowering they will repeat in September. The doubles

Monarda didyma *(3)*

flower from June through to Aug. **G. *p. Albiflorum*.** Pure white form of the meadow cranesbill. Single flowers. June-July. Ht. 60cm. **G. *p. Midnight Reiter*.** Deep purple leaves. May-June. Violet blue single flowers. Very compact form. Ht. 30cm. Sp. 25cm. **G. *p. Mrs Kendal Clarke*** has delicious, scrubbed blue single flowers Ht. 60cm. Sp. 30cm. **G. *pratense Summer Skies*** has tight double, pale blue and white flowers. Ht. 65cm. Sp. 30cm. **G. *p. Victor Reiter*.** This is the original black leafed G.pratense. A much better doer. Leaves as they mature turn reddish green. Violet single flowers June-July. Ht. 30cm. Sp. 40cm. **G. *p. Violacea Flore Pleno*** has rich purple, fully double flowers. Very lovely. Ht. 65cm. Sp. 30cm. **G. *psilostemon*.** E. A. Bowles says that this plant, although 'flaunting that awful form of floral original sin, magenta', is redeemed by its magnificent, black eye. June to Sept. It is interesting that Reginald Farrer, a contemporary of Bowles, shared this detestation of magenta. Robert Gathorne Hardy makes some illuminating comments on the history of magentaphobia in his book 'Three Acres And A Mill', published in 1939. Reginald Farrer, he writes, "was a product of the aesthetic nineties; and one of the fashionable fancies of that period was to despise many shades of purple under the vilifying term of magenta. Since that time it has been possible for people with no taste of their own to utter what they think may be a safe and proper opinion, by professing their dislike for any flower which may possibly be called magenta. I have even known that most exquisite of all our flowers, the little autumn blooming Cyclamen neapolitanum, to be rejected with its scarcely less beautiful winter successor, Cyclamen coum, by a man who said: "No, I don't like them; they're magenta." I dislike such dogmas as this, which remind me of the dogmas of Fascism and Communism and extreme Pacificism, and extreme Toryism. Their devotees require no thought for solving a problem, but merely refer to a sort of fallacy-ridden intellectual logarithm table, and finding the answer there, utter arrogantly their not-to-be-controverted opinion." Seventy five years have elapsed since the publication of Robert Gathorne Hardy's book, but magentaphobia remains rife (with the honourable exception of Christopher Lloyd) and horticultural bigotry is still the worst weed in the garden. G. psilostemon has very large, cut leaves, which in autumn turn rusty red. Ht. 120cm. Sp. 70cm. Flower stems need some support. Sun or part shade. Any soil. Perhaps the best of all geraniums in regards foliage. Christopher Lloyd suggests planting with Amsonia. **G. *renardii*** is a compact, hardy Cranesbill. Neat dome of foliage, similar to sage in colour and texture. White flowers with bold violet veining. June-July. Immaculate front of border plant. Ht. 25cm. Sp. 30cm. **G. *x riversleaianum Mavis Simpson*** has wonderful, soft pink flowers. Christopher Lloyd calls it "one of the finest clamberers and interweavers without being in any way aggressive." May-Oct. Very decorative silver green foliage. A cracker! Dislikes heavy clay. Ht. 20cm. Sp. 40cm. **G *x r. Russell Pritchard*.** Spreading carpet. Bright, magenta pink flowers. Silver green foliage. May-Oct. Ht. 40cm. Sp. 60cm. G. S. Thomas suggests planting with blue or white Agapanthus. **G. *Rozanne*** is a wonderful G. wallichianum hybrid. Huge intense blue flowers with white centres. June-Nov. Better in every way than G. Buxtons Variety Ht 60cm. Sp. 100cm. **G. *Sabani Blue*.** Deep blue flowers. Very early flowering. May-June. A new G. ibericum hybrid. Ht. 40cm. Sp. 30cm. **G. *Salome*.** Large violet flowers with a

Nigella damascena *(3)*

black eye. June-Oct. Yellowish leaves. Ht. 25cm. Sp. 55cm. ***Geranium sanguineum*** is a British native, and has been in cultivation for over three hundred years. We know it was one of the flowers Gerard grew in his garden at Holborn. Its English name is 'The Bloody Cranesbill', but as Alice Coate writes, "Whoever saw blood of such a flaring magenta?" I have vivid memories of this native plant growing in the rocks of Sutton Bank. Ascending the steep, twisting road, my mother, who was driving, lifts both hands off the driving wheel and exclaims in delight. Bloody indeed it nearly was. G. sanguineum, although thriving in any well drained garden soil, in the wild is nearly always found on limestone, hence Sutton Bank and the magnificent colonies on the limestone paving in the Burren, Co. Clare. The species, with its magenta flowers. June-Sept. **G. *s. Ankums Pride*** has fluorescent pink flowers. June-Sept. **G. *s. Album*,** has pure white, flowers. June-July. **G. s. *Max Frei*.** Light purple flowers. June-July. Notable for making a neat rounded hummock. Good autumn colour. **G. *s. Striatum*** (Queen Anne's Needlework). Discovered growing wild in the sand on the Lancashire coast, was introduced into cultivation in the early 17th Century. It has light pink flowers, veined with red, and flowers from June to September. All varieties have green, much divided leaves and a ground hugging, prostrate habit. They need full sun and thrive in any well-drained soil, including chalk. All are capable of making extensive mats, up to 50cm. in diameter. Ht. 20cm. **G. *Shocking Blue*.** A wallichianum hybrid. Large, bright blue flowers with white centres above green, mottled leaves. June-Nov. Ht. 20cm. Sp. 30cm. **G. *Sirak*** (gracile x ibericum). Striking large mauve flowers. June-September. Ht. 40cm. Sp. 30cm. Likes moist cool soil. **G. *Spinners*.** Large violet purple flowers. May-July. Ht. 60cm. Sp. 30cm. **G. *Spring Fling*.** Introduced by Bob Brown. Spectacular spring foliage. Cream, yellow and green, flushed with pink and with brown markings. Foliage turns green in summer. No reversion. Pink flowers May-Oct. Ht. 35cm. Sp. 30cm. **G. *Sue Crug*.** Vivid pink flowers with a black eye. May-Oct. Yellowish leaves. Ht. 35cm. Sp. 50cm. Sun. Any well drained soil. **G. *Terre Franche*** (G. platypetalum x G. renardii). Terre Franche forms a mound of quilted sage-green leaves. Large mid blue flowers veined with purple. June-July. Ht. 35cm. Sp. 30cm. Good with yellow coreopsis. **G. *versicolor*.** Spreading Hardy Geranium. Evergreen blotched foliage. White trumpet flowers with red veins. May-October. Ht. 30cm. Sp. 40cm. Tastes do vary. I think G. versicolor a lovely dainty thing. Christopher Lloyd thinks it "rather horrid and weedy looking." **G. *wallichianum Buxton's Variety*** is always a plant in short supply. It can only be propagated from seed. Flowers throughout late summer and autumn. Blue flowers with white centres. Needs good rich loam to do well. For the most intense blue should be planted in dappled shade. Needs to be given plenty of space as it produces long trailing stems. Ht. 30cm. Sp. 60cm. **G. *wlassovianum*** has greyish, hairy leaves. Pinkish violet heavily veined flowers. Good autumn colour. Sun or part shade. June-August. Ht. 30cm. Sp. 60cm. **G. *Zetterlung*.** Another Geranium renardii hybrid. White flowers veined with violet. May-July. Puckered sage green leaves. Ht. 25cm. Sp 30cm. All renardii hybrids are for me to be cherished. Their neat mounds of foliage look superb planted in bold groups in gravel; but do leave enough space between the plants to be able to appreciate their exquisite form.

Oenothera speciosa *(2)*

Hardy Geraniums for Shade

Many of the Geraniums listed in our section on sun loving plants will also be successful in part shade. We recommend in particular G. palmatum. Listed below are Geraniums, which are 100% successful in deep shade. **G. *macrorrhizum*** is an excellent evergreen, ground cover plant. It has strongly apple scented foliage and spreads vigorously. Flowers May-Sept. Ht. 20cm. Sp. 50cm. **G. *m. Czakor***, has magenta flowers, and **G. *m. Ingwersens Variety***, is pale pink. **G. *m. Spessart.*** White flowers. **G. *nodosum*** is a native woodlander. Small lilac flowers. May-Sept. Glossy leaves. A fine foliage plant for deep shade. Ht. 25cm. Sp. 35cm. **G. *phaeum*** is native to the Balkans. It has been in cultivation for over three hundred years; Gerard grew it in his famous garden at Holborn. G. phaeum is a hardy evergreen plant, which thrives in areas of the most impossible dry shade. It is vigorous and spreads by means of a rhizomous root; it also reliably self-seeds. It has handsome reflexed flowers on 60cm. stems. Sp. 60cm. If, after it finishes flowering in June, it is cut to the ground, it will respond with fresh basal foliage and a second flowering in September. The species, for the darkness of its flowers, bears the English name, "The Mourning Widow", but as Jason Hill writes; "the name is not very appropriate, for it has not only the light elegance, which is common to all the Geraniums, but its flared back petals and curling stamens give it an elvish, fantastic charm, which suggests a widow who has ceased to mourn." Christopher Lloyd is predictably snooty about Geranium phaeum, calling it "a miserable thing, that does not show up at all." My mother once had a tiny, shady town garden given over almost entirely to blue Forget-me-nots and the black flowers of Geranium phaeum. Very restful to the eye it was. But then Christopher Lloyd only ever praises loud clamouring plants. Of the numerous forms of G. phaeum, I like **G. *phaeum Samobor***, collected in the Balkans by Elizabeth Strangman, which has the dark purple flowers of the species and dramatic chocolate splashed leaves and **G. *p. Album*** which has glistening, pure white flowers. **G. *sylvaticum*.** Sylvaticum means woodland, which is this plant's native habitat, although in cultivation it seems to perform equally well in full sun. Native to Scotland, it is not as drought tolerant as G. phaeum and prefers reasonably fertile soil. It has handsome, deeply divided, clean green foliage and flowering stems about 60cm. tall. Sp. 30cm. It flowers throughout May and June. Non-invasive. **G. *s. Mayflower*** has clearest blue, white centred flowers, **G. *s. Bakers Pink***, has shell pink flowers with a white centre, **G. *s. Album*** is pure white, and the quite lovely, **G. *s. Amy Doncaster***, has the darkest blue, white throated flowers. All are deliciously fresh and crisp, both in flower and foliage.

Origanum dictamnus *(3)*

Grasses

I must confess to being a latecomer to the appreciation of Grasses. I think this largely stems from having seen them as the exclusive property of those uninspiring garden designers, who limit themselves to dwarf conifers, grasses and heathers and eschew all manner of flowers, being terrified that a surfeit of colour might undermine their own suburban weariness! But of course that is a thing of the past since the advent of Piet Oudolf. Piet Oudolf's particular influence has been in the use of some of the larger grasses with bold, late flowering perennials. He uses grasses such as Calamagrostis, Sorghastrum, Deschampsia, Molinia, Stipas, and Miscanthus with Achilleas, Aster horizontalis, Campanula lactiflora, Eupatoriums, Filipendula, Foeniculeum vulgare Purpureum, Heleniums, Hemerocallis, Lythrums, Monardas, Sanguisorba, Sedums and Thalictrums to create bold exuberant borders. Go visit Piat Oudolf's wonderful garden at Hummelo in Holland, or if, like me, you are short of time, buy Piet Oudolf's and Michael King's inspirational book 'Gardening with Grasses', published by Frances Lincoln.

Grasses for Sun

ACORUS. (Acoraceae). The Acorus family are not strictly grasses, (they belong to the same family as the Arums and Zantedeschias), but because of their neat fans of grass-like foliage, they are, like Ophiopogons, (which are in fact members of the lily family), used by gardeners for grass like foliage effects. ***Acorus gramineus*** is a North American native, which thrives in moist rich soil. It is also at home in the bog garden and will grow in shallow water up to 20cm. deep. William Robinson recommends it for use as an edging plant in the basins of fountains. It is stone hardy and its foliage remains fresh into winter. Acorus should be grown in full sun. The foliage and root, which contain the essential oil Calamus, when bruised, emit a warm aromatic scent. It has rather dull, arum-like flowers, which are normally only produced when grown in water. Both the following varieties grow about 20cm. tall with a spread of 25cm. They are excellent subjects for pots and display their best leaf colour in winter. **A. *g. Ogon*** has fans of bright yellow and green striped leaves. **A. *g. Variegatus*** has silver and green striped foliage

BRIZA. (Poaceae). **B. *media*** is evergreen and flowers June-Aug., bearing 60cm. panicles of locket shaped white flowers. Sp. 50cm. Best in sun, but will grow in part shade The flower heads are long-lasting and persist into late summer. They dangle most elegantly. Because of the lightness of its flower heads, it needs planting in largish groups. Briza media is a very popular plant with the dried flower brigade; but this should not be held against it. It is a cool season grower.

BROMUS. (Poaceae). **B. *inermis Skinners Gold.*** Perennial grass. Native to Europe. This form has leaves with broad green margins and a light yellow central stripe. Light yellow flowering stems add to the effect. June-July. Sun. Ht. 100cm. Sp. 30cm.

Paeonia tenuifolia *(3)*

CALAMAGROSTIS. (Poaceae). **C.** ***x acutiflora Karl Foerster***, one of the classic plants of modern landscape design, is chiefly valued for its strong vertical lines. Makes an excellent subject for massed plantings. Pinkish grey flowers in June. Cool season grower. **C.** ***x a. Overdam*** is a sport of Calamagrostis Karl Foerster. It has white and green, striped foliage tinged with soft pink. The foliage tends to become tired and lose its colour, but this grass responds well to being cut back in mid season, promptly producing a new flush of brightly coloured foliage. Both plants will grow in any soil including heavy clay. Sterile, so no troublesome seedlings. Ht. 100cm. Sp. 50cm.

CAREX. (Cyperaceae). Carexes or sedges are indigenous to cooler, temperate regions. They grow happily under most conditions, including as specimens in pots. Most dislike drought. **C.** ***buchananii*** makes 50cm. high clumps of evergreen, coppery brown leaves. Sp. 45cm. The tips of the leaves are curled like little pigs' tails. It is distinct from other New Zealand brown grasses in having a distinctly upright habit. It needs good drainage and tolerates drought, once established. A native of New Zealand, it is not utterly hardy, and its crowns should be protected with a loose mulch in winter. **C.** ***phyllocephala Sparkler*** is the brightest of the variegated sedges with rich, dark green leaves, neatly striped with creamy white. Untypically for a sedge, the leaves are borne in whorls on the tops of erect 60cm. stems. Sp. 30cm. Best in shade, it is a warm season grower and needs rich moist soil. White negligible flowers in October.

CHIONOCHLOA. (Poaceae). **C.** ***rubra***, the New Zealand red tussock grass has olive green, arching stems. 90cm. tall. Spread 30cm. Perhaps the most handsome of all the New Zealand grasses with its uncluttered light airy habit.

CORTADERIA. (Poaceae). **C.** ***selloana Pumila*** is a useful dwarf (Ht. 120cm. Sp. 80cm) with spikes of white flowers. Aug-Nov. Can look good if planted with other mighty beasts – Phormiums make a good contrast.

DESCHAMPSIA. (Poaceae). **D.** ***cespitosa Goldtau***. A compact form with flowerheads which mature to a warm gold colour. June-July. Ht. 60cm. Sp. 90cm. Moist soil. Sun. Best planted in large groups. Good for naturalising. **D.** ***c. Northern Lights***. A dramatic, new variety. White and green striped leaves. Bronze flowers. Ht. 30cm. Sp. 30cm. Good in containers. **D.** ***flexuosa Tatra Gold*** is grown for its neat rosettes of needle like golden foliage. The bright yellow of the leaves only slightly fades in the summer. Bronze flowers in June. Prefers an acid soil and will perish on chalk. Slow growing. Looks good with Uncinia uncinata. Sun or shade.

FESTUCA. (Poaceae). The fescues are cool season growers. They are all drought tolerant and will thrive on any well-drained soil. In rich soil they tend to lose their neat, compact character. Young plants have the best colour, so annual division in the spring is advised. If left undivided the foliage will become less intense in colour and the plant will gradually rot from the centre. **F.** ***glauca Elijah Blue*** as inanimate as a blue plastic pin cushion. For those who like that kind of thing. Plant it in a sea of crushed blue glass interrupted by lumps of slate (as depicted on a cover of The Garden Magazine). A plague on designers!

Papaver Patty's Plum *(2)*

IMPERATA. (Poaceae). **I. *cylindrica Rubra.*** Stunning Japanese grass. Emerges in spring with red tips and yellowish green leaves. By mid summer the whole leaf has turned rich blood red. Best planted where the blades can catch the low evening light. Roger Grounds describes the effect as like a glass of red wine held up to the light. A warm season grower, it flourishes in fertile moist soil and full sun. Described by many as hardy, it is definitely tender in our frosty garden. It makes an elegant specimen in a pot, stood outside a west window to catch the evening sun. Plants in pots survive quite happily under the staging in a cold greenhouse in winter. Ht. 40cm. Sp. 50cm. Looks splendid with Geranium Jolly Bee.
LIRIOPE. (Asparagaceae). I have avoided Liriopes up to now. Too neat as a garden plant for my Libertarian tastes. They are however, as I have discovered, rather splendid in containers. Look good all the year round and require no special attention. The fresh, rich, evergreen foliage arches elegantly over the side of a pot; the flowers are a minor amusement, - although picked they merit more attention. **L. *muscari Ingwersen*** has violet blue flowers. Aug-Oct. Ht. 25cm. Sp. 25cm. **L. *m. Munroe White*** has white flowers. July-Sept. Ht. 30cm. Sp. 25cm.
MISCANTHUS. (Poaceae). **M. *sinensis.*** I am keen on their flowers, especially as seen by moonlight, their strong upright lines and their good autumn colour. Although normally described as plants for richer soils, all Miscanthus will grow happily in even the driest of sandy soil, once established. Miscanthus will not show their true character till about three years after planting, and are best left undisturbed for many years. They all flower Sept-Oct. Their flowers remain decorative through the winter. **M. *s. Kleine Silberspinne*** is a dwarf variety, which has narrow leaves; its flowers open reddish and then turn silver. Ht. 140cm. Sp. 80cm. It makes a good specimen in a large pot. Slightly larger are **M. *s. Ferner Osten,*** and **M. *s. China,*** both bred by the great Ernst Pagels. M. s. China he considered to be the finest of his offspring. It has narrow, olive green leaves, large red flower heads, and its foliage colours brilliantly in autumn – Ht. 160cm. Sp. 60cm. M.s. Ferner Osten is a delicate looking plant, which does not suffer from the somewhat coarse looks of some of its family. Very deep red flowers. Very early flowering. Ht. 150cm. Sp. 100cm. **M. *s. Strictus*** is an improved form of M. s. Zebrinus, the Tiger grass, to which it is identical in appearance although much tougher and more reliable. Its plain green spring foliage, by summer, has become horizontally striped with green. It looks straight out of a painting by Henri Rousseau and at any moment a tiger might leap from under its foliage. Handsome with Salvia involucrata Bethellii and deep red dahlias. Ht. 270cm. Sp. 80cm. **M. *s. Variegatus*** has crisp, vertically striped, green and white leaves. Slower to clump up than most of the M. sinensis varieties. Ht. 180cm. Sp. 80cm. **M. *s. Morning Light*** has very narrow leaves with a cream variegation. Turns a lovely pinky buff in autumn. An elegant plant, which does well in dry soil. It seldom flowers in this country. Ht. 80cm. Sp. 60cm. A graceful plant for a large container. **M. *s. Cosmopolitan*** is a dramatic American hybrid with very broad white and green striped foliage. Makes a handsome specimen plant. Ht. 300cm. Sp. 100cm. **M. *s. Flamingo,*** whose dark pink flowers retain their colour well into autumn. It has a distinctive arching habit. Ht. 200cm. Sp. 100cm. **M. *s. Gracillimus.*** Fine leafed.

Paradisea liliastrum *(3)*

Makes a graceful rounded plant. Very late flowering. Red flowers Oct. Golden yellow foliage in the fall. Ht. 210cm. Sp. 60cm. **M. s. *Silberfeder*** is the tallest Miscanthus. Once established it soars to 250cm. Sp. 100cm., flaunting aloft its silver pennants.

MOLINIA. (Poaceae). **M. *caerulea.*** Miscanthus are stalwart, somewhat masculine plants. Molinias are altogether more feminine. Although, like Miscanthus they make strong vertical lines, their stems are much lighter and their flower heads more airy. Like Miscanthus, Molinias provide good autumn colour and remain attractive in a skeletal kind of way through winter. **Shorter Molinias. M. *c. Variegata*** has neat rosettes of smartly variegated cream and green foliage and produces purplish spikes of flowers on stems of the most exquisite pale buttery yellow in early summer. In autumn the whole plant turns brown. Aug-Sept. Ht. 40cm. Sp. 35cm. **M. *c. Moorhexe*** makes handsome compact rosettes of dark, steely green, spike like leaves. In early August it produces erect spikes of striking black flower heads. Ht. 100cm. Sp. 40cm. In autumn its fiery orange foliage and flowering stems torch through the darkest of November mists. **M. *c. Heidebraut.*** Produces a mass of slender flowering stems with purple flowers in Aug-Sept. The whole plant becomes a blaze of yellow in autumn. Ht. 110cm. Sp. 60cm. **M. *c. Strahlenquelle.*** The stems of this grass arch outwards. Narrow panicles of purple flowers in Aug-Sept. Ht. 90cm. Sp. 30cm. **Taller Molinias. M. *c. ssp. arundinacea Karl Foerster*** forms an erect plant with open, delicate flower heads. Aug-Sept. Ht. 200cm. Sp. 60cm. **M. *c. ssp a. Transparent*** grows to 240cm. Sp. 60cm. and is perhaps the best of all. Perfectly upright, it bears the most delicate open panicles of dark brown flowers in September. Weighed down by rain these will gracefully droop to the ground, only to rise erect, when dry again. M. Transparent persists well through the winter and still retains the delicate structure of its flowering stems and the warmth of its soft yellow, autumn colour well into January.

OPHIOPOGON. (Asparagaceae). Although technically a member of the lily family, Ophiopogon is often, because of its appearance, classified as a grass. **O. *jaburan Vittatus*** has neatly striped silver and green leaves. Ghostly white flowers in July-October. O. j. Vittatus is superficially like Liriope Silver Dragon but makes good dense clumps, unlike the latter which has a rather sparse stoloniferous habit. Ht. 15cm. Sp. 20cm. **O. *japonicus Sea Mist*** is a miniature with very fine thread-like silver and green leaves. Ht. 8cm. Sp. 15cm. Good in a trough. **O. *planiscapus Nigrescens*** bears small, insignificant, blue flowers in August, which are followed by handsome, black seed heads. It is grown for its handsome evergreen deep black, strap like leaves. Planted against bare soil, Ophiopogon fails to make an impression. It makes an excellent subject for a pot, its black foliage, being handsomely offset by a terracotta rim. Also looks good in gravel. For another effective contrast follow Christopher Lloyd's suggestion and plant with the deep green fern Blechnum penna-marina. For spring interest underplant with snowdrops or scillas. Moist rich soil. Sun or shade. Stone hardy. A warm season grower. Ht. 15cm. Sp. 20cm.

Passiflora caerulea *(3)*

PANICUM. (Poaceae). P. ***virgatum*** . Switch grass. Bears panicles of tiny flowers July-Sept. These create a light misty effect. The thin ribbon like leaves turn red in autumn. Non-invasive and clump forming. Allow a spread of at least 40cm. The tall, airy, flowering stems are a wonderful yeast for autumn bouquets; they hold up well though the winter and make a lovely armature for the frost. A warm season grower. **P. *v. Heavy Metal*** is notable for its bluish green foliage, irregularly streaked with purple. Panicles of pink grey flowers. Ht. 80cm. **P. *v. Rotstrahlbusch*.** Selected for its red autumn leaf colour. Brown flowers. Ht. 120cm. **P. *v. Squaw*** is distinguished by the pink tones in the flowers. Some red tints appear in foliage in autumn. Ht. 120cm. Sp. **P. *v. Warrior*** has large, airy, brown flowerheads. Ht. 80cm. **P. *v. Strictum*.** Blue green leaves. Open panicles of brown flowers. Taller than any other Panicum. I like my grasses big so this one appeals to me. Grows, I am told, to 180cm. high.

PENNISETUM. (Poaceae). Some of the North American forms of P. alopecuroides do not flower well in the English climate. Both the following varieties are reliable. **P. *alopecuroides Woodside*** bears purple brown bottlebrush flowers in Sept-Oct. Ht. 60cm. **P. *a. Hameln*** is a reliably flowering slightly shorter form, whose flowers open white, later turning brown. Ht. 50cm. Sp. 40cm. **P. *orientale*** bears airy pink flowers from June onwards through to Sept. In autumn the flowers become an attractive smoky grey. Shorter than P. alopecuroides, it grows to 35m. in height. Sp. 35cm. Looks well planted against a background of Allium Sphaerocephalum. Only hardy in well-drained soil. Self seeds in a restrained kind of way. **P. *setaceum Rubrum*.** Extremely architectural. The whole plant, leaves and stems and flowers are a rich burgundy colour. Flowers Aug-Nov. Unfortunately will not cope with temperatures below freezing so must be brought in in winter. Tolerates extreme drought. Magnificent in large containers. Superb with Eucomis Sparkling Burgundy. Ht. 75cm. Sp. 40cm. **P. *villosum*.** Comes from the mountains of N. East Africa. Pure white flowers. June-Nov. Good as both a cut flower and in the garden. Best in full sun and moist soil. Normally hardy in England. Very occasionally gets killed by frost, but reliable self seeder and its seedlings are always in flower by mid summer. Very showy in containers. Ht. 100cm. Sp. 50cm.

PLEIOBLASTUS. (Poaceae). These dwarf bamboos are excellent in small tubs or pots. **P. *variegatus*.** Dwarf bamboo with neatly striped green and white leaves. Ht. 50cm. Sp. 35cm. **P. *viridistriatus*.** Dwarf bamboo with bright yellow and green leaves. Ht. 75cm. Sp. 35cm.

SASA. (Poaceae). **S. *veitchii*.** The edges of this plant's rich green leaves dry out in summer to a parchment colour, giving the leaves the appearance of being variegated, an effect which persists through the winter. Ht. 70cm. Sp. 30cm. Shade.

SESLERIA. (Poaceae). **S. *caerulea*.** Native to Britain. Thrives on poor chalky soil. Leaves glaucous blue on upperside, green on the underside. Small panicles of black flowers July-Aug. Ht. 30cm. Sp. 30cm. Good for naturalising. Sun.

Pelargonium Black Knight *(2)*

SORGHASTRUM. (Poaceae). **S. *nutans Indian Steel*** has blue green foliage. Its flowers are borne in reddish brown spikelets with yellow anthers in July-Aug. In autumn the whole plant turns red. Ht. 140cm. Sp. 40cm. A strong upright plant. Non-invasive. A warm season grower.

SORGHUM. (Poaceae). **S. *Purple Majesty*.** Showy purple leaves with a red midrib down the center of each leaf. Bristly beige seedheads. Dramatic North American annual. Ht. 140cm. Sp. 70cm.

SPARTINA. (Poaceae). **S. *Pectinata*.** Prairie Cord Grass. Long arching yellow and green striped leaves, which turn orange in autumn. Black flowers. Invasive rhizomes. Sun or part shade. Any soil. Ht. 150cm. Sp. 100cm.

STIPA. (Poaceae). All varieties of Stipa grow in the wild in open positions and dislike being crowded in the herbaceous border. They all (except Stipa arundinacea) thrive in poor dry soil and conserve moisture by rolling up their leaves to reduce transpiration during drought. **S. *arundinacea* syn. Anemanthele lessoniana.** The Pheasant Grass has delicate, narrow, arching leaves, which become spotted with orange and red in autumn. It bears long, trailing panicles of brown flowers in July-Aug. Not a drought lover. Ht. 45cm. Sp. 40cm. **S. *brachytricha*** (there is some confusion over its name; it also bears the names Calamogrostis brachytricha and Acnatherum brachytricha) has plain green leaves, which are often bronze flushed, both in spring and again in autumn. It bears large plumes of delicate, feathery flowers August-September, which persist into the winter and provide a wonderful vehicle for the frost. Will grow in sun or part shade. Makes a large clump. Ht. 120cm. Sp. 80cm. **S. *calamogrostis*** has a long flowering period, from June-Oct. Stiff, bluish green leaves. Arching stems topped by large, pale gold flower heads. June-July. Looks good in winter. Ht. 60cm. Sp. 30cm. Much loved by Piet Oudolf. Makes a good specimen plant. **S. *capillata*** has greyish green foliage and forms an erect plant with wispy flower heads. July. Ht. 60cm. Sp. 30cm. One of the freshest looking grasses for early spring. **S. *gigantea*** makes not overtly impressive mounds of greyish green foliage from which erupt in July 200cm. stems, which dangle spikelets of golden oat like flowers. Sp. 80cm. If I could have only one grass in my garden this would be it. Looks handsome right through to the end of October. Good as a feature on a corner. **S. *tenuissima*** has fine light green foliage. Arching heads of fluffy flowers in May-July. When planted in groups, the flowers delicately billow. Ht. 35cm. Sp. 25cm. Looks good into November. All Stipas are cool season growers.

UNCINIA. (Cyperaceae). **U. *uncinata*.** Mahogany red sedge. Looks good in gravel. Makes a wonderful specimen in a pot. Needs rich, moist soil and protection in cold winters - so we are told, although so far it has proved perfectly hardy, here, in our chilly frost pocket. A cool season grower. Ht. 20cm. Sp. 30cm. Negligible brown flowers in August.

UNIOLA. (Poaceae). **U. *latifolia* syn. Chasmanthium latifolium** has broad bladed leaves. Flowers hang in curious panicles, which have the appearance of being ironed flat. June-Aug. The flowers change from green to coppery pink in autumn. Needs moist rich soil. A cool season grower. Ht. 60cm. Sp. 35cm.

Pelargonium echinatum *(2)*

Grasses for Shade

CAREX. (Cyperaceae). **C. *dolichostachya Kaga Nishiki*** makes a fountain-like mound of golden margined green leaves. Evergreen. Long lived and very hardy. Prefers moist soil in part shade. Ht. 60cm. Sp. 60cm. **C. *elata Aurea* syn. Carex stricta Aurea**, Bowles Golden sedge, was discovered by E A Bowles as a boy, when naturalising in a Norfolk marsh. The most handsome of all the sedges. Tall yellow leaves with green stripes. Elegant brown flowers in June-July. Needs to be moist. Ht. 60cm. Sp. 60cm. Will grow in water. Best grown in part shade, as the variegation tends to fade in full sun. Non-invasive. **C. *grayi*.** Decorative flowerheads shaped like a Norman mace - as used by Bishop Otto of Bordeaux in the Bayeux Tapestry. Being a Bishop, by Canon Law, he was not supposed to shed blood, but to crush not to pierce was alright. And if blood squeezed out? Well that was not his intention. So the Bishop, no doubt, still went to Heaven. May-July. Ht. 60cm. Sp. 30cm. Moist soil. Shade. **C. *pendula Moonraker*.** Striking creamy yellow variegated foliage. Colour at its best in winter. Graceful pendulous flowers April-June. Discovered in a hedge in Wiltshire. Ht. 100cm. Sp. 60cm. Woodland plant. **C. *siderosticha Variegata*** is a brightly brazen, erect plant, resembling a small Hosta. The green leaves are neatly edged with white. Thrives in moist shade or bog. Ht. 25cm. Sp. 35cm. Beware of slugs.

HAKONECHLOA. (Poaceae). **H. *macra Alboaurea*.** Roger Grounds calls this the most enchanting, golden grass ever introduced. A slow, warm season grower, Hakonechloa is an ideal plant for container growing. Its narrow, yellow leaves are held in lax, arching fans. It likes rich, moist soil and will grow in sun, but is happiest in cool shade. At Wisley they used to plant it with the steely blue Hosta Halcyon. Ht. 15cm. Sp. 30cm.

LUZULAS. (Juncaceae). Woodrushes are versatile evergreen plants, which will grow in sun or deep shade and thrive in dry or wet soil. **L. *nivea*,** the snowy woodrush, makes neat clumps of strap like, deep green leaves, edged with showy white hairs. Evergreen and smart in winter. Flat heads of creamy white flowers, which are excellent for cutting to give substance to mixed bunches of spring flowers. April-June. Ht. 45cm. Sp. 30cm. **L. *sylvatica*,** the greater woodrush, forms bold rosettes of succulent like foliage. It makes effective ground cover even in the driest of shade. As well as the species there are two named forms: **L. *s. Aurea*** (previously wrongly named L. s. Hohe Tatra), whose pale green summer leaves turn bright yellow in winter and **L. *s. Marginata*,** which has rich green leaves, neatly edged with cream. All varieties of L. sylvatica grow about 60cm. tall with a spread of 100cm. They are bomb proof plants for tubs or large pots, in which they look good at all times of the year, surviving all manner of neglect. Ideal container plants for those who find wielding a watering can fatiguing. All Luzulas are cool season growers.

MILIUM. (Poaceae). **M. *effusum aureum*.** Bowles Golden Grass. Yellow stems, leaves and flowers. Wonderful for lighting a dark corner. Superb with the Mourning Widow Cranesbill or Brunnera macrophylla. A cool season grower, it fades away in the heat of summer. May-July. Ht. 60cm. Sp. 40cm.

Pelargonium Lord Bute *(2)*

Herbaceous Plants for Shade

ACONITUM. (Ranunculaceae). This plant gained one of its English common names, Monkshood, from the hooded shape of its flowers. Aconite's other common English name, Wolfsbane, derives from its extreme toxicity. As Ann Pratt tells us: 'it is well known, that wolves and other wild animals have been killed by arrows dipped in the juice of aconite.' Gerard writes 'of their very fair and goodly blew flowers, in shape like an helmet, which are so beautiful, that a man would thinke they were of some excellent virtue, but non est semper fides habenda fronti.' Turner says of Aconitum that it is a "most hastie poison." The flowers of Aconites are so perfectly designed to be pollinated by bumble bees that a plaster cast taken from the inside of the flower forms an almost exact replica of a bumble bee of medium size. Monkshood is a persistent plant, found on the site of vanished gardens. It thrives in dark shady places and resents drought. Christopher Lloyd is predictably no great fan of Aconites – they are not sufficiently strident in colour for the Sage of Great Dixter. He quotes approvingly John Raven for saying that aconites "always strike me as dingy and lifeless", compared with Delphiniums and Veronicas. In case we are still in any doubt to his feelings on Aconites, Christopher Lloyd goes on to say "that all aconites look in need of a good wash." but then he always was one for gaudy baubles. **A. *napellus vulgare Albidum*** has creamy white flowers. June-Aug. Ht. 90cm. Sp. 30cm. **A. *cammarum Bicolor*** has white flowers with a picotee blue edge. July-Aug. Ht. 90cm. Sp. 30cm. **A. *septentrionale Ivorine*** has elegant, pale creamy flowers. June-Aug. Ht. 50cm. Sp. 30cm. **A. *henryi Sparks Variety*** has deep blue flowers. July-Aug. Ht. 120cm. Sp. 30cm. **A. *carmichaelii Arendsii*.** Huge deep blue flowers. Sept-Oct. Bold architectural foliage. Sturdy upright habit. For me quite one of the best late flowering plants. Looks wonderful with dahlia David Howard. Ht. 80cm. Sp. 40cm. **A. *c Royal Flush*** has spectacular deep red juvenile foliage in spring, and flowers which we are told are darker than those of Arendsii. Sept-Oct. Ht. 60cm. Sp. 30cm. **A. *Stainless Steel*.** Greyish blue flowers with cream throats. June-Aug. Ht. 80cm. Sp. 30cm.

AEGOPODIUM. (Apiaceae). **A. *podagraria Variegata*.** This is a much less aggressive form of that noxious common weed, Groundelder, or as the Scots would have it, Bishopsweed because like the Bishops it engrosses all. The English name 'Jump About' graphically illustrates its mobility! The variegated form is much less invasive and a very pretty weed for dry shade. In flower it is particularly fetching, its flat heads of white flowers tastefully matching the bold white variegation of its foliage. It is a plant much employed by arrangers of subtle bouquets. And if you get a fit of gout at the wedding reception, Aegopodium will reputedly alleviate the pain! The leaves boiled also make a passable substitute for spinach. June-August. Ht. 60cm. Sp. 80cm.

AJUGA. (Lamiaceae). **A. *reptans*.** Our native bugle is a "blacke herbe and groweth in shadowy places and moyst groundes." William Turner called it black because of its foliage. Its flowers are a transcendent blue. Bugle also goes by the name of thunder and lightning, presumably because of the threatening dark storm clouds of its foliage. All Ajugas flower April-May. **A. *r. Arctic Fox*.** Leaves are tricoloured: small grey centre surrounded by an expanse of cream with a narrow edge of dark green. Bright blue

Pelargonium Francis James *(2)*

flowers. Ht. 15cm. Sp 30cm. Bob Brown rates this variegated Ajuga very highly. For my taste a bit flashy. **A. *r. Braunherz*** has particularly intense blue flowers Ht. 15cm. Sp. 80cm. **A. *r. Catlins Giant*** is bigger in all its parts. Ht. 25cm. Sp. 80cm. John Raven in his book A Botanist's Garden, recounts how he underplanted Ajuga with Anemone fulgens and Tulipa sprengeri, both of whose scarlet flowers were shown off to advantage by the black foliage of the Ajuga. **A. *r. Chocolate Chip*** is a form of Bugle with a very compact habit. A welcome characteristic in a genus, which is normally rampageous. Small black leaves. Blue flowers May-July. Ht. 25cm. Sp. 25cm. **A. *reptans Golden Beauty*** has blue flowers May-June. Gold and green foliage. We are told A. Golden Beauty overwinters better than A. Arctic Fox. Ht. 15cm. Sp. 30cm.
ALCHEMILLA. (Rosaceae). **A. *mollis***, or Lady's Mantle, is one of the freshest of ground cover plants for sun or part shade. It has large, soft green leaves and frothy sprays of lime green flowers. June-Sept. Wonderful picked in mixed bunches of cottage garden flowers. Ht. 40cm. Sp. 60cm. **A. *erythropoda*** has small, scalloped, blue grey leaves with correspondingly small, lime green flowers. June-Sept. Ht. 15cm. Sp. 25cm. Super neat plant for those who find A. mollis too unruly.
ANAPHALIS. (Asteraceae). **A. *triplinervis.*** One of the most silver of silver leafed plants. Compact habit. White flowers. July-Sept. Native to China. According to Beth Chatto, grows by rocky streams in The Himalayas alongside Persicaria. Unlike most silver leafed plants, will thrive in both heavy soil and shade. Looks delicious planted with pink Japanese Anemones. A favourite among the dried flower brigade. Ht. 45cm. Sp. 30cm.
ANCHUSA. (Boraginaceae). The name Anchusa derives from the Greek for 'paint for the skin'. Like borage, to whose family they belong, the Anchusas have intense blue flowers. **A. *sempervirens*** (also known as Pentaglottis sempervirens and in plain English as Alkanet) is a thug; it will grow in any soil, including the poorest and driest and once established is impossible to eradicate. Anchusa sempervirens seems to have been a mediaeval introduction and is often to be found near abbeys, perhaps grown for the red dye yielded by its roots. It is a great persister. The Tradescants planted it in their garden in the early 17th Century and it was one of the few original plants remaining in 1749. Anchusa sempervirens, unlike Anchusa azurea, tolerates deep shade and is good for naturalising in semi-wild areas. April-July. Ht. 90cm. Sp. 120cm.
ANEMONE. (Ranunculaceae). As Geoffrey Grigson remarks, the English name Windflower, which is a direct translation from the Greek "Anemone", is most apt for "the flowers do hang and nod and shake in the wind". ***A. nemerosa*** is native to our woodlands and if found growing wild in a wood is a strong indicator that the wood is of ancient origin. "Moll o' The Woods" (its Somerset vernacular name) grows from slowly creeping rhizomes, which eventually form a large clump. The foliage and flowers push up with mesmerizing speed in early March when its brilliant white flowers light up the woodland floor. In a cool spot the flowers will linger into May. The wood anemone's foliage dies back in early summer and the whole plant goes dormant until the following spring. Although native to our woodlands, it will grow in an open position. It likes humus rich, but well drained soil. Plants need to be marked carefully so as to avoid digging up the small rhizomes during the dormant season. Ht. 20cm. Sp. 20cm.

Pelargonium Lord Roberts *(2)*

A. ***n. Robinsoniana,*** a choice variety with large lavender blue flowers. A. n. Robinsoniana is perhaps my favourite plant in the spring garden for both delicacy of its colouring and its form. Ht. 20cm. Sp. 20cm. **A.** ***japonica*** despite its misleading name is native to China not Japan. Live plants were first introduced into England by Robert Fortune in 1844. It was soon crossed with the Nepalese white flowered species A. vitifolia and from this cross have been produced the modern varieties, which are now correctly called **A.** ***x hybrida.*** This is a plant tolerant of all except the driest soil but will thrive best in a heavy loam. All varieties flower from August to October. The largest flowers are obtained by planting in dappled shade, although this may delay flowering until September. They all should be allowed a spread of at least 40cm. **A.** ***x h. Honorine Jobert*** (introduced in 1851) has glistening white flowers. Ht. 80cm. **A.** ***x h. Hadspen Abundance*** has elegant, small, cupped pink flowers. Ht. 70cm. **A.** ***x h. Richard Ahrens*** has large light pink flowers. Ht. 100cm. **A.** ***x h. Pamina*** has deep red double flowers. Ht. 70cm. **A.** ***rivularis.*** Native to North India. White flowers with blue reverse and blue anthers. May-June. Ht. 38cm. Sp. 20cm. Likes rich moist soil. **A.** ***sylvestris.*** Native to Europe. Fragrant nodding white flowers in profusion from May-July and odd flowers thereafter through into the late autumn. The ripe seed heads, like lumps of cotton wool, are a very white and decorative. Tends to run freely in lighter soils while in heavier soils not such a gad-about. Ht. 35cm. Sp. 30cm.

AQUILEGIAS. (Ranunculaceae). **A.** ***canadensis.*** A native of Canada and North Eastern America where it grows in woods and shady places. Long narrow flowers with red spurs and yellow petals. May-June. Ht. 30cm. Sp. 20cm. The following varieties are all hybrids of our native **A.** ***vulgaris*** and are as happy in shade as in sun. Flowers will last longer in shade and plants will be considerably taller. All varieties grow to between 60 and 90cm. high, depending on situation and soil. They flower May-June. **A.** ***v. Black Barlow*** has double spurless, blackish purple flowers. **A** ***v. Norah Barlow*** has double, spurless, cream, green and purple flowers - very Victorian looking. **A.** ***v. Rich Purple*** has spurless, double, very rich, dark purple flowers. **A.** ***v. Variegata*** has green foliage splashed with yellow. Variable single flowers - blue, pink or white. **A.** ***v. William Guiness*** has inky black and white flowers. **A.** ***v. Alba*** has ivory white flowers and glaucous leaves. The species **A.** ***vulgaris*** has variable blue and white bicoloured flowers. All look lovely planted in bold swathes.

ARUM. (Araceae). **A.** ***italicum Pictum.*** A selected form of the native Lords and Ladies; Arum italicum Pictum has wondrously marbled foliage, which provides handsome winter ground cover. Summer dormant. Indestructible in dry shade. A plant which, because of its morphology - spadix in the spathe, has in the English vernacular attracted a bewildering myriad of sexually suggestive names: Priests Pintle, Dog's Dibble, Red hot poker, Schoolmaster, Cow's Parsnip, Snake's meat.... and, I fear, Lords and Ladies are no better than they should be. Ht. 15cm. Sp. 25cm.

ARUNCUS. (Rosaceae). All varieties thrive in damp, shady conditions. **A.** ***aethusifolius*** comes from Korea and makes compact clumps of fresh green, finely cut

Pelargonium Mystery *(2)*

leaves, which turn red in autumn. Abundant spikes of small cream, Astilbe-like flowers in June. Decorative brown seedheads. Ht. 30cm. Sp .30cm. **A. *dioicus*** or Goat's Beard has an astonishing geographical range, being native to North America, Europe, The Caucasus, China and Japan. It has large plumes of creamy white flowers in July-Aug. Light green leaves. Ht. 120cm. Sp. 60cm. A. ***d. Kneiffii*** has dainty, thread like leaves. It produces open panicles of Astilbe like, cream flowers in July-Aug. Ht. 90cm. Sp. 50cm.

ASARUM. (Aristolochiaceae). **A. *splendens*** has large cyclamen-like evergreen leaves, marbled with silver. Curious brown and white banded flowers from February to June. Ht. 17cm. Sp. 25cm. Flowers are borne at soil level and fertilised by slugs! Somebody ought to tell hard line feminists. It just shows - there is always somebody worse off than yourself.

ASTILBE. (Saxifragaceae). The word Astilbe derives from the Greek and translates as 'lack of shine' - enough to make any plant feel a little inferior. They are however useful plants for the margins of streams and ponds or any damp soil. Their flower heads although without shine, make plumes, which remain handsome, even when gone to seed. **A. *x japonica Deutschland.*** White flowers. June-July. Ht. 50cm. Sp. 40cm. **A. *x j. Red Sentinel.*** Dark red flowers. July-Aug. Ht. 50cm. Sp. 50cm. **A. *x arendsii Etna.*** Dark red flowers. July-Aug. Ht. 60cm. Sp. 50cm. **A. *simplicifolia Sprite.*** Shell pink flowers in July-August. Dark foliage. Ht. 20cm. Sp. 50cm. I am no great lover of dwarf Astilbes but this is a cracker. Delicate, elegant and well proportioned! **A. *s. Willie Buchanan.*** Creamy white flowers. July-Aug. Ht. 30cm. Sp. 50cm. **A. *thunbergii. Professor Van Der Wielen*** is an excellent architectural plant with arching heads of ivory flowers August-September. Ht. 110cm. Sp. 120cm.

ASTRANTIA. (Apiaceae). The name Astrantia is probably a corruption of Magisterantia, hence its English name, Masterwort. Astrantia was introduced from Southern Europe and grown as a supposed antidote to poisons and as a cure "for all pestilentiall carbuncles and blotches, and such other great apostemations and swellings". (Gerard). It is a useful plant for not too dry shade, where its flowers are remarkably long lasting. **A. *major*** has greenish white flowers and will grow in poor dry sand. July-Sept. Ht. 60cm. Sp. 25cm. **A. *m. Buckland*** has green tipped white outer ray petals filled with tiny pink flowers. Very free flowering. Ht. 75cm. Sp. 25cm. **A. *m. Celtic Star*** has very large white flowers – 9cm. across – a good cut flower. May-October. Ht. 60cm. Sp. 30cm. The spectacular deep red form, **A. m. Hadspens Blood**, is listed in the section for sun loving perennials; it needs strong light to develop its best colour.

BERGENIA. (Saxifragaceae). "Seekers after no trouble ground cover, with landscape architects at the van, fawn upon bergenias". (Christopher Lloyd). Used in excess bergenias can be ponderous. I rather sympathise with John Raven who wrote: "Goodness knows why, but the insufferably coarse genus of Bergenia is apparently coming back into favour. If you grow Bergenias for the elegance of their foliage you can have no appreciation of elegance and if for their flowers you must be colour blind." Like many such statements however it needs qualification. **B. *Abendglut*** has semi double shocking magenta pink flowers April-June. Crinkle edged leaves which turn red in winter. Ht. 25cm. Sp. 50cm.

Pelargonium Pink Capricorn *(2)*

B. ***Bressingham White*** may have unremarkable foliage but it has very acceptable white flowers in April-May. Ht. 30cm. Sp. 60cm. **B. *Eroica* syn. B. Overture** has tawdry, bright rose pink flowers (best perhaps decapitated with the garden shears) but in winter its green foliage colours the most resplendent bronze purple – wonderful with the low winter sun shining through. Tastes do differ though – I have seen this plant's winter foliage compared (by one who neither admires Bergenias, nor enjoys consuming offal) to slices of stale, raw liver! Ht. 40cm. Sp. 50cm. **B. *Morgenrote*.** Deep pink flowers on long stems April -June. Repeat flowers in in autumn Ht. 60cm. Sp. 60cm. **B. *Rosi Klose*** bears heads ofclear pink flowers on tall stems. Floriferous Ht. 30cm. Sp. 50cm. **B. *Silberlicht*.** Palest pink flowers. Ht. 30cm. Sp. 40cm. Bergenias will grow in any soil and are happy in sun or shade, although B. Eroica and B. Abendglut will have the best winter colour when planted in full sun. They are generally trouble free plants, although vine weevil can be a problem and in sluggy gardens Bergenias' fleshy leaves must be protected with garlic pellets.

BLECHNUM. (Blechnaceae). **B. *penna marina*.** Evergreen fern New Zealand hardy ground cover. Narrow deep green fronds. Ht. 20cm. Sp. 40cm. Moist or dry soil. Christopher Lloyd recommends it for planting in paving in cool shade. **B. *spicant*.** Evergreen fern. Narrow lance shaped fronds. Glossy, dark green leaves. Ht. 35cm. Sp 40cm. Dry or moist, not too alkaline soil. Shade or part shade.

BLETILLA. (Orchidaceae). **B. *striata*** comes from China. It has pleated, sword shaped leaves and purple orchid flowers. June-July. Ht. 40cm. Sp. 25cm. It will grow in sun or shade and thrives in moist humus rich soil. I like the very chaste white form **B. *striata Alba*.**

BRUNNERA. (Boraginaceae). Brunneras are named after the Swiss 19th Century botanist Samuel Brunner. It was discovered in 1800 growing in woods North of Tiflis. Native to Eastern Europe and Western Siberia, they are extremely hardy. They grow best in shade and need a rich moist soil, which does not dry out in summer. Forget me not flowers, April-June, are followed by large, heart shaped leaves. Ht. 55cm. Sp. 70cm. The flowers are elegant and shown to advantage without any foliage. The summer foliage provides dramatic ground cover. **B. *macrophylla*. *Betty Bowring*** has pure white flowers and deep green leaves, **B. *m. Langtrees***, a vigorous variety has silver grey spotting on its foliage and blue flowers, **B. *m. Jack Frost*** is definitely the best Brunnera yet - dramatic silver leaves with dark green veins and a dark green margin. Usual blue flowers March-May. Needs full shade or the leaves will lose their brightness and **B. *m. Hadspens Cream*** an easy variegated form with pale green and cream foliage. Blue flowers. Not nearly so apt to burn as Dawson's White.

CARDAMINE. (Brassicaeceae). ***C. pratensis*** or Lady's Smock is a much loved native plant. Gerard wrote "it comes out for the most part in April and May, when the cuckoo begins to sing her pleasant song without stammering." It is called the Ladies Smock because "they are very abundant in the moist part of the meadows and at a distance look like large white patches, resembling the inner female garment hung out to bleach". (Thornton 1810). Delicious single pale mauve flowers April-May. The plant for all its modest delicacy bears in some parts of the country the thoroughly indecent name "cuckoo pintle." ***C. pratensis Flore Pleno*** is a delicious double form, which is sterile and spreads by little bulbils, which form on the underside of the leaves. Cardamine pratensis needs moist or boggy soil and part shade. Ht. 30cm. Sp. 30cm.

Penstemon Suttons Pink Bedder *(2)*

CIMICIFUGA. (Ranunculaceae) **syn. Actea**. Bugbane. A member of the buttercup family. All the bugbanes like a cool moist soil and thrive in rich woodland conditions. They are all hardy. **C. *racemosa*.** Scented, white flowers. July-Aug. Green leaves. Ht. 125cm. Sp. 25cm. **C. *racemosa Brunette*** has the most elegant, dark bronze, divided foliage and bears pure white, bottlebrush flowers, which have a strong orange blossom fragrance. Despite its height never needs staking. It is slow to increase. July-Sept. Ht. 90cm. Sp. 25cm. **C. *r. Hillside Beauty*.** We are told the blackest form yet. Scented pink flushed white flowers July-Sept. Ht. 100cm. Sp. 35cm. **C. *r. James Compton*** is shorter than C. r. Brunette and has really black foliage. White scented flowers July-Sept. Ht. 80cm. Sp. 25cm. **C. *r. Pink Spike*.** A form with soft pink scented flower spikes. July-Sept. Near black foliage. Ht. 125cm. Sp. 25cm. **C. *simplex White Pearl*,** native to Japan and Mongolia; it has white, bottlebrush flowers. Oct-Nov. Deeply divided green foliage. Ht. 120cm. Sp. 35cm.

CONVALLARIA. (Asparagaceae). **C. *majalis*.** "Lily of the Valley". A native of Britain and Northern Europe. As Alice Coats tells us "In pagan times it was considered the special flower of Ostara, the Norse godess of the dawn". Unusually for such a wide spread wild flower the classical authors were "silent as fishes" (Brunfels) on the subject of its medicinal properties; however by the late 17th Century lily of the valley water was used to alleviate a wide range of ailments, "it having an especial property to help weak memories, raise Apopletick persons, cheer the heart, and ease the pain of gout". (Rea. 1665). Lilies of the Valley were introduced into English gardens in the early 16th Century. Scented white flowers in May. Ht. 15cm. Sp. 25cm. Moist, cool, woodland conditions.

CORYDALIS. (Papaveraceae). C. ***flexuosa*.** Japanese woodlander related to the Dicentras. The species has china blue flowers and green leaves and a form which goes by the name of **C. *f. Purple Leaf*,** which has identical flowers but deep bronze foliage. Both are excellent plants for moist shade. Summer dormant, they have delightful fresh winter foliage followed by sheets of scented blue flowers. April-June. Good in pots but need annual splitting and repotting every autumn to keep their vigour. Ht. 20cm. Sp. 35cm.

CRINUM. (Amaryllidaceae). **C. *x powellii*.** Although Crinum comes from the Greek for lily they are in fact members of the Amaryllis family. Crinum x Powellii is an easy, majestic plant. It bears bold heads of fragrant, pink flowers on 120cm. stems and makes handsome mounds of lush green, strap like leaves. It needs moisture and rich feeding. Books tend to make a fuss of this plant, demanding full sun and a protective winter mulch for its successful cultivation. We find it stone hardy and much easier to cultivate in shade, where its unquenchable thirst for water is easier to satisfy. I have seen it happily wading out into a pond, where it unfailingly flowers in the shade of a poplar. Drought, though, it will not tolerate. I grow the very beautiful pure white form, **C. *x. p. Album*.** Crinums look magnificent grown in tubs. Unlike lilies they are free from all diseases and pests. Aug-Sept. Ht. 100cm. Sp. 100cm.

CRYPTOTAENIA. (Apiaceae). **C. *japonica Atropurpurea*.** A very tolerant plant for dry shade. Purple foliage. Upright habit. Insignificant flowers. Ht. 46cm. Sp. 25cm. Self seeds to form bold swathes. Looks good in late summer and early autumn when most shade plants look weary.

Phlomis fruticosa *(6)*

CYCLAMEN. (Primulaceae). The name derives from 'cyclos', the Greek for circle, probably a reference to the spiral coil into which the flowering stem converts as the seed heads mature. Cyclamen does have an ancient English common name "Sowbread" which is derived according to Lindley from it being, despite its acridity, the principal food of the wild boars of Sicily. **C. *hederifolium*** was valued in this country as a medicinal root long before it was grown as a garden plant. The classical author, Apuleius suggests: "in case that a man's hair fall off take this same wort and put it into the nostrils". In Tudor times it was considered an aphrodisiac, "Being beaten and made up into... little flat cakes, it is reported to be a good amorous medicine to make one in love, if it be inwardly taken." Gerard grew ten species and varieties of Cyclamen at Holborn. John Hill wrote in 1757, "A very little Trouble and Attention will serve to stock a Garden with this humble Plant, but there will require that Patience which should be as much the Characteristic of the Gardener as of the Angler; for it will be some years before seedlings rise to flower." I like Vita Sackville West's description of Cyclamen hederifolium, "the little frightened cyclamen, with leveret ears laid back". **C. *coum*.** Vivid pink flowers. Jan-March. Plain dark green leaves with crimson undersides. Ht. 12cm. Sp. 25cm. Shade or part shade. Any well drained soil. Drought tolerant. Summer dormant. **C. *c. album*** White flowers. Jan-March. Plain dark green leaves with crimson undersides. Ht. 12cm. Sp. 25cm. Shade or part shade. Any well drained soil. Drought tolerant. Summer dormant. **C. *hederifolium Album*.** White flowers in Sept-Oct. Marbled foliage. Summer dormant. Good in dry shade under trees. Ht. 10cm. Sp. 15cm. **C. *persicum*.** The wild ancestor of our winter houseplants has all the grace and freshness that they do not. White or pale pink twisted reflexed flowers. Tender – greenhouse or conservatoty. Febuary. Ht. 25cm. Sp. 20cm.

DIANELLA. (Xanthorrhoeaceae). **D. *tasmanica*.** Quite why the goddess Diana got saddled with this one nobody knows. Dianellas are closely related to Phormiums. Like Phormiums, Dianellas have stiff, sword shaped leaves but more modest in scale. Ht. 45cm. Sp. 25cm. In July they have starry blue flowers, which are followed by dramatic, dark blue berries. It is for the berries this plant is primarily grown. Native to Tasmania, Dianellas are reasonably hardy and do well in moist peaty soil in a sheltered, partly shaded position.

DICENTRA. (Papaveraceae). The Greek name refers to the twin spurs on each flower. Dicentras are a member of the fumitory family and closely related to Corydalis. Dicentra spectabilis was first introduced into this country from China in 1810. It quickly became popular as a garden and greenhouse plant. Although seldom grown now as a greenhouse plant it does make an excellent subject for forcing in the early spring. Dicentra spectabilis, being dormant in summer, will tolerate more drought than Dicentra formosa, although it makes a more impressive plant with some moisture. Dicentra formosa, like Corydalis flexuosa, needs a humus rich woodland soil. All Dicentras' main flowering period is May-June, but they will continue to put out odd flowers in July. **D. *spectabilis*.** As well as the traditional Pink form of "Lady in the Bath" I like the quite ravishing and unblushing, pure white version, **D. *s. Alba*.** Both flower

Physostegia virginiana *(4)*

May-July. Ht. 90cm. Sp. 35cm. ***D. formosa***. All varieties flower May-July and have a height of about 30cm. and a similar spread. **D. *f. Aurora*** has greyish blue foliage topped by pure white flowers, **D. *f. Bacchanal*** has deep red flowers and **D. *f. Langtrees*** has bluish foliage and white flowers. **D. *King Of Hearts*** is a hybrid. Bob Brown rates it the best Dicentra yet. Its parentage involves D. peregrina, D. formosa subsp. oregana and D. eximea. Robust, compact form. Blueish feathery leaves. Deep pink flowers May-Nov. Ht. 18cm. **D. *Ivory Heart***. White flowers April-Aug. Silvery blue leaves. Ht. 30cm. Very crisp looking.

DIGITALIS. (Plantaginaceae). Foxgloves gained their botanical name from 'digitabulum', the Latin word for thimble. Our common name is supposedly a corruption of Folk's glove or Fairy's glove. Some see the wild foxglove as a sinister plant. Maeterlinck describes it as shooting up "like a melancholy rocket". Digitalis is still used as a therapeutic drug for cardiac problems. Foxglove stems were apparently used in Kent as parasol handles. **D. *purpurea*** is native to our woods and hedgerows. It prefers sandy or gravelly soils and can be hard to establish on heavy land. Although biennial, it quickly becomes resident in gardens, self-seeding liberally, if you are not too eager with the secateurs and allow the seed heads to mature. One plant can produce a million seeds! For those tasteful beings, who require all their self-seeded foxgloves to be white, it is possible at a juvenile stage to rogue out all the purple seedlings as they may be distinguished by a faint flush of purple in the leaf stalk. Two varieties of Digitalis purpurea: **D. *p. Alba***, the very lovely white form and **D. *p. Apricot***, which is not apricot at all but a voluptuous pink. Both flower May-July. Ht. 140cm. Sp. 30cm.

DODECATHEON (Primulaceae). Dodecatheons are native to North America. Dodecatheon meadia was first introduced into England by Bishop Compton, who grew it at his famous garden in Fulham at the beginning of the 18th Century. Each flower has four or five petals, which are joined at the base and sharply reflexed – very similar in form to Cyclamen, although the petals of Dodecatheons tend to be proportionally longer. They are bulbous plants, which like a cool, well drained, humus rich and not too dry spot. Like Cyclamen they are supremely elegant. The great William Robinson wrote in his 'Alpine Flowers' that the Dodecatheon "is second to none of our old border-flowers. Its blooms should be seen in early summer in every spot worthy of the name of a garden." **D. *meadia Album*** has white flowers May-June. Ht 40cm. Sp. 25cm. **D. *m. Queen Victoria*** has lilac rose flowers May-June. Ht. 50cm. Sp. 25cm.

DORONICUM. (Asteraceae). D orientale was first introduced from the Caucasus into England in 1815. **D. *orientale Finesse*.** Leopards Bane. This selection has yellow daisy flowers, with much finer petals and longer flowering stems than the species. Hairy heart shaped leaves. Ht. 35cm. Sp. 30cm. Flowers from early April through to the end of May.

EPIMEDIUM. (Berberidaceae). Epimedium translates from the Latin as a bishop's mitre and describes the flower shape of these elegant woodland plants, whose flowers have great delicacy. They thrive in any humus enriched soil, which does not become bone dry in summer. Their evergreen foliage provides good ground cover. The old foliage should be cut off in early spring before flowering so as not to obscure the flowers. The

Plumbago capensis *(1)*

following varieties, all of which are about 20cm. tall, with a spread of about 25cm, **E. *pinnatum Colchicum*** - yellow flowers veined with red, **E. x *versicolour Sulphureum*** - delicate, sulphur yellow flowers and delightful juvenile foliage, delicately shaded with bronze, **E. x *youngianum Niveum*** - starry white flowers and chocolate coloured, juvenile foliage, ***Epimedium* x *youngianium Roseum*** - lilac pink flowers and chocolate juvenile foliage **E. x *warleyense*** - orange flowers and **E. *x perralchicum*** - bright yellow flowers.
EUPATORIUM. (Asteraceae). The genus takes its name from King Eupator of Pontus, who used one species as an antidote to poison. **E. *purpureum*** comes from the Eastern states of America, where it is called Joe Pye Weed. G. S. Thomas calls it one of the most imposing of herbaceous plants. It has robust, dark purple stems, lanceolate foliage and flat heads of pinkish purple flowers. Although preferring rich, moist soil, it does surprisingly well in my impoverished sand. Ht. 160cm. Sp. 100cm. **E. *rugosum* syn. Ageratina altissima** comes from the North East of America and has flat heads of white flowers in late summer. A useful plant, it thrives in the driest of dry shade, where we plant it with Helianthus Lemon Queen, which against all the rules thrives under the same conditions. Ht. 110cm. Sp. 100cm. **E. *r. Chocolate*** has an identical habit and flowers. Chocolate foliage. ***E. ligustrinum* syn. Ageratina ligustrina** is, unlike E. purpureum and E. rugosum, an evergreen shrub. It has glossy, mid green, lanceolate foliage - very clean and fresh looking - and cream heads of flowers with a rose pappus. In Mexico, where it originates, it grows to 450cm. In this country, it seldom makes more than 120cm. in height with a spread of 100cm. For a sheltered spot. Evergreen and untouched by frost in mild winters.
EUPHORBIA. (Euphorbiaceae). **E. *amygdaloides robbiae*** is an essential plant for deep shade. On 60cm. stems it bears whorls of dark evergreen foliage, topped by yellow flowers in spring. Spreads stoloniferously and therefore good for ground cover. Few ground cover plants look so smart and have such architectural qualities. Will grow in sun or shade. Thrives in clay. Looks starved in sand. Drought tolerant. April-May. Sp. 80cm.
FILIPENDULA. (Rosaceae). **F. *rubra Venusta*.** This plant comes from the damp meadows of North Eastern America and bears the very American name "Queen of the Prairies." It sends up great spires of 200cm. branches, well furnished with large jagged leaves and topped with deep rose pink, Astilbe like flowers. A classic plant for the bog garden. G. S. Thomas recommends planting it in boggy woodland with Rhododendrons to lighten the oppression of the latter's heavy evergreen foliage. Sp. 150cm.
GALIUM. (Rubiaceae). **G. *odoratum* syn. Asperula odorata.** Woodruff is a native to British woodlands. White flowers May-June. As Ann Pratt says the small white flowers "seem firm and compact, as if cut out of wax." Its Old English name 'wudorofe', perhaps derives from its roving habit, gently spreading with a running root. Woodruff develops a sweet smell of hay as it dries, hence its use as a strewing herb and to stuff mattresses and to sweeten linen. Woodruff tea was used as a remedy for fevers, colds and Consumption. It was according to Gerard put into wine "to make a man merrie", but any claim for its use as a stimulant seems to be spurious. Its dried flowers however make a delicious tisane. G. odoratum thrives in humus rich soil. Ht. 25cm. Sp. 35cm.

Polemonium caeruleum *(3)*

GUNNERA. (Gunneraceae). G. ***manicata***. Native to Colombia. White flowers. July-Aug. Huge leaves. Ht. 200cm. Sp. 200cm. Moist but not waterlogged soil. Good in clay. Does best in a cool position. G. ***tinctoria*** (native to Chile) is similar but slightly smaller in all its parts; its leaves are more frilly than G. manicata. In winter cut off the leaves and pile them up on top of the crown to give protection from frost. "I believe the great secret for ensuring its reaching gigantic proportions is to feed the brute" (E. A. Bowles). The leaves of Gunneras are not fully developed till midsummer. This allows them to be underplanted with spring flowering perennials such as Hellebores and primroses

HELIANTHUS. (Asteraceae). Helianthus are listed in the section on sun loving plants. What else would one expect of members of the sunflower family? But **H. *Lemon Queen*** also grows superbly in dry deep shade. Tall late flowering perennials for dry deep shade are a rarity. Try planting it with Eupatorium rugosum.

HELLEBORUS. (Ranunculaceae). Hellebores light up the winter and early spring like no other plants. **H. *foetidus***, the Stinking Hellebore, "a great branching plant" as Gilbert White describes it, is the first to come into flower, lighting up our gardens in January with its luminous, lime green flowers. It grows in the poorest of sandy soil and is a useful plant for dry shade. A native to this country, it was once grown, as Gilbert White tells us, as a drug to cure children of worms, but, as he says, it had somewhat unpredictable results; "Where it killed not the patient it would certainly kill the worms; but the worst of it is, it will sometimes kill both." The delights of herbal medicine! H. foetidus makes a large plant and can grow to 90cm. in height. Sp. 60cm. It makes plentiful seed, whose oily coating is caviar to the slug. Slugs only eat the coating and not the seeds themselves, which are expelled in the excremental slime, which coats their bodies, and thus distributed around the garden. (Slugs have their uses! - for further benefits see Aspidistra). **H. *orientalis***, the Lenten Rose, is often in flower in my garden before the Christmas Rose! It is a variable plant and grown from seed can have flowers from deepest plum to pure white. Ht. 60cm. Sp. 25cm. H. orientalis is by nature a woodlander. Plant with Primulas, Corydalis flexuosa and Pulmonarias. ***H. o. Atrorubens*** has purplish flowers. Exceptionally early flowering. Always in flower before Christmas. We have listed H. niger and H. corsicus as sun loving plants.

HEPATICA. (Ranunculaceae). Hepaticas are alpine woodland plants and like a leafy soil and partial shade. They are lime lovers and a useful trick is to plant them between pieces of limestone. The limestone will also keep their roots cool and provide a marker to stop you digging them up when they are dormant. They are slow growing plants and care must be taken in a garden situation to insure they are not overwhelmed by more vigorous plants. Hepaticas also make choice specimens for the alpine house. If grown in pots, they should be plunged in a shaded cold frame during the summer. Hepaticas hate disturbance and should only reluctantly be lifted or divided. Propagation is best by fresh seed sown immediately in a cold frame. **H. *nobilis*.** Violet blue flowers with showy white stamens. March-April. Deep green basal leaves. Ht. 10cm. Sp. 15cm. Moist soil. Part shade. Slow growing. **H. *transylvanica*** is a more robust plant than H. nobilis. Febuary-April. Ht. 15cm. Sp. 20cm.

Primula Gold Lace *(2)*

HEUCHERA. (Saxifragaceae). Heucheras are American members of the saxifrage family. They provide good, evergreen ground cover in not too dry soil. They flower from May till the end of June and are happy in either full sun or part shade. I am no admirer of dark leafed Heucheras. To me they look distinctly artificial like silk flowers manufactured in some huge Shanghai factory to furnish the gardens of crematoriums all over the world. **H. *Caramel*.** Pale caramel leaves. White flowers. Much stronger grower than Amber Waves. Ht. 25cm. Sp. 25cm. **H. *Chocolate Ruffles*** has chocolate ruffled leaves, which are maroon on the undersides. It has sprays of small white flowers. Ht. 25cm. Sp. 25cm. **H. *Obsidian*.** The blackest yet. Ivory flowers on red stems. Ht. 30cm. **H. *Plum Pudding*** has very deep red leaves. Tiny maroon and green flowers. Ht. 45cm. Sp. 45cm. **H. *Silver Scrolls*** has silver leaves with red purple veins and purple underside. Ht. 30cm. Sp. 35cm. **H. *Venus*** has heavily silvered leaves with deep green veins. Buff flowers. Ht. 55cm. Sp. 30cm.

HOUTTUYNIA. (Saururaceae). **H. *cordata*.** Leaves and roots scented, says G. S. Thomas of Seville oranges - to others like me they are utterly nauseating. Christopher Lloyd says the leaves can be eaten. Good ground cover for moist soil. Perhaps looks best grown in shallow water. Good for masking the edge of a pond liner. Spreading. All varieties grow to 30cm. with a spread of 60cm. and are June-September flowering. **H. *c. Flore Pleno*,** has showy, double white flowers. Its foliage colours well in the autumn. **H. *c. Chameleon*,** a flamboyant foliage plant with leaves variegated green, bright red and yellow. By comparison, its modest, small white flowers appear sartorially challenged.

KIRENGESHOMA. (Hydrangeaceae). **K. *palmata*.** Black stems bear voluptuous fat buds, which open into elegant pendant soft yellow bells. Aug-Sept. Elegant. Shade. Must have moist fertile soil. Despite what the books say, lime tolerant. Ht. 85cm. Sp. 60cm.

LAMIUM. (Lamiaceae). Useful ground cover plants for light not too dry shade. Welcome for early spring flowers. Both varieties flower April-June and grow about 20cm. in height and 40cm. in spread. Lovely for picking for small spring posies with Pulmonarias and Primulas. **L. *maculatum Pink Pewter*.** Shell pink flowers and heavily silvered leaves. **L. *m. White Nancy*** has almost completely silver leaves and ivory white flowers.

LIBERTIA. (Iridaceae). Libertias are named after Marie Libert the 19th Century student of liverworts - there's an inspiring subject! **L. *grandiflora*** is a New Zealand relative of the Sisyrinchium family. Panicles of snow white flowers. May-June. Orange seedpods, which split open, to show shiny black seeds. Sheaves of sword-like dark green leaves. Ht. 60cm. Sp. 50cm.Can be planted in Shade or Sun. Well drained soil. An architectural plant, which is a good deal more robust than its Sisyrinchium cousins. **L. *peregrinans*.** Makes a smaller plant and is distinguished by the broad orange band on its light green leaves. White flowers May-Sept. Ht. 40cm. Sp. 30cm.

LIGULARIA. (Asteraceae). Ligularia derives its name from ligula, the Latin for a little tongue, a reference to the tongue shaped ray florets. Ligularias are members of the daisy family and need moist soil or bog conditions. They grow in sun or part shade, are summer flowering, and need protecting with slug pellets. **L. *clivorum Britt-Marie Crawford*** is a good selection with much blacker leaves and stems than older clones such

Rudbeckia maxima *(6)*

as L. c. Desdemona. Ht. 100cm. Sp. 120cm. **L. *hessei Greynog's Gold*.** Large orange yellow daisy flowers. July-Sept. The biggest of the Ligularias. Ht. 175cm. Sp. 100cm. **L. *palmatiloba*** has huge, palmate leaves and shaggy, yellow flowers. Ht. 120cm. Sp. 120cm. **L. *przewalskii*,** has elegant, almost black stems, basal foliage of dark green deeply cut leaves and narrow spires of small clear yellow flowers. Ht. 120cm. Sp. 30cm. **L. *stenocephala The Rocket*.** Large triangular leaves with jagged edge. Spikes of orange yellow flowers on purplish stems. Ht. 120cm. Sp. 100cm. Moist soil.

LILIUM. (Liliaceae). Most lilies of course need lots of sun and fertile soil. I have a garden with large areas of dry sandy shade. One of the few plants, which grows successfully in these areas is **L. *philipinense*.** Large white flowers, handsomely streaked with crimson on the reverse. August-September. No need of staking and - the big surprise - in my well-drained soil apparently stone hardy. The stems are clothed with narrow dark green foliage. They remain handsome even when the pendant trumpets are replaced by large upraised seedpods. I have them planted along a woodland path where they take over duty in summer from swathes of spring foxgloves. Ht. 120cm. Sp. 15cm.

LUNARIA. (Brassicaceae). Purple Honesty. The latin name is a reference to the moon-like appearance of the beautiful silver seed pods. Purple flowers April-May. An annual but a willing self seeder. Ht. 90cm. Sp. 25cm. Tolerates a wide range of conditions. Happy in dry shade.

LYSICHITUM. (Araceae). **L. *camtschatcensis*** is a smaller Japanese version of the N. American Lysichitum. Unlike the North American, yellow flowered plant, which bears the vernacular name of skunk cabbage, L. camtschatcensis has sweetly scented white flowers. May-June. Moist cool acid conditions. Large banana-like, peagreen leaves Ht. 40cm. Sp. 60cm.

LYSIMACHIA. (Primulaceae). The Greek word 'lysis' means loose, and 'machia' means strife. Lysimachia placed under the yoke was supposed to calm strife between draft horses and oxen harnessed to the same plough. **L. *ciliata Firecracker*.** Rather obvious lemon yellow flowers. April-Aug. Redeemed by splendid mahogany foliage. Ht. 60cm. Sp. 60cm. **L. *clethroides*** is a useful plant for its bright red shoots in spring, and its late summer flowering. Native to China and Japan, like L. ephemerum, it needs a moist soil to do well. Its white flower heads have an attractive arching habit. Ht. 100cm. Sp. 80cm. Unlike L. ephemerum, L. clethroides can be invasive; it looks well naturalised by a stream. **L. *ephemerum*.** Elegant stems of waxy, grey foliage terminate in long spires of starry white flowers. Flowering in August and September, it makes a fine companion for Salvia uliginosa. It thrives in moist fertile soil, but in dry sandy gardens will visibly pine. John Treasure apparently grew it at Burford House in shallow water. Lysimachia ephemerum, unlike Lysimachia punctata, is non-invasive. Although native to Spain, it is stone hardy. Lysimachia ephemerum is one of those leggy beauties, which are much prized by arrangers of tasteful bouquets. It grows to 100cm. and never needs staking. Sp. 60cm. Lysimachia punctata, the common yellow loosestrife abounds on our native river banks. It has brazen yellow flowers, coarse foliage and a bullying invasive root. **L. *p. Alexander*** is a pretty form with elegant grey and white variegated foliage. Useful on the edge of a natural pond. Perhaps the most attractive feature of L. punctata is its sexy, deep red buds pushing through the soil in early spring. Dreary yellow flowers July-Aug. Ht. 75cm. Sp. 60cm.

Sanguinaria canadensis *(3)*

LYTHRUM. (Lythraceae). Lythrum's common English name is Purple Loosestrife. This leads to much confusion with Lysimachias; Lysimachia punctata is known as Yellow Loosestrife. To avoid such confusion I prefer to call Lythrum by one of its old dialect names: Long Purples. John Clare so called it, and it is a graphic and beautiful name, conjuring up those tall handsome spires, standing proud on our riverbanks. Lythrum salicaria literally means the willow like Lythrum. This is commonly taken as referring to the plant's narrow willow leafed foliage. Though perhaps fanciful, I find more pleasing Ann Pratt's suggestion, that it gained its name from being so often found among willows. Although a native of the river bank, **L. *salicaria*** does not insist on boggy conditions and is an easy robust plant for the moist border. It flowers throughout August and September, a period when the garden is not over-rich in colour. It is happy in part shade or sun. **L. *s. Blush*.** Spires of pale rose pink flowers. Ht. 120cm. Sp. 60cm. **L. *s. Zigeunerblut*.** "Gypsy Blood." Deep red flowers. Ht. 120cm. Sp. 60cm.
MACLEAYA. (Papaveraceae). Plume Poppy. Named after Alexander Macleay, a 19th Century Australian Colonial official, the Macleaya, is a native of East Asia and a member of the poppy family. It is a bold architectural plant, well clothed with huge, jagged, grey-green leaves. Ht. 180cm. Sp. 90cm. It needs moist fertile soil, to perform well and thrives in a cool North border. **M. *cordata Flamingo***, has creamy pink flowers and **M. *microcarpa Kelways Coral Plume***, has smaller deep coral pink flowers. Both varieties flower July-Sept. M. cordata is normally preferred for the border as it has a non invasive habit. M. micocarpa with its aggressive running roots is ideal for the wild garden.
MATTEUCCIA. (Woodsiaceae). **M. *struthiopteris*.** Fronds of pale yellow-green foliage form an upright chalice, hence its English name, the Shuttlecock Fern. Moist shade. Ht. 90cm. Sp. 70cm.
MECONOPSIS. (Papaveraceae). **M. *sheldonii Lingholm*.** Himalayan blue poppy. July. Mid green hairy leaves. Ht. 100cm. Sp. 60cm. Dappled shade. Moist, leafy, acid soil. A cool position. M. s. Lingholm is more perennial and slightly more tolerant of dry conditions than the M. betonicifolia.
MERTENSIA. (Boraginaceae). **M. *virginica*.** The Virginian Cowslip is according to Gertrude Jekyll "the very embodiment of the freshness of early spring." Smooth grey leaves. Arching sprays of blue flowers. April-May. It likes cool woodland conditions and dies down in mid summer. Ht. 40cm. Sp. 30cm. Vita Sackville West recalls showing a Mertensia virginica plant to American visitors to her garden at Sissinghurst. The Americans taking this plant for granted as a wildflower were closed off to its beauty and were amazed she gave it garden room. Such is the snobbery of gardeners!
MYOSOTIDIUM. (Boraginaceae). **M. *hortensia*.** Chatham Island Forget-me-not. Large sprays of blue flowers in July. Handsome glaucous foliage. Ht 45cm. Sp. 45cm. Light shade. Moist, fertile soil. Good in frost-free maritime gardens.
NEPETA. (Lamiaceae). **N. *subsessilis*.** All the other Nepetas are listed in the section for sun loving plants but this Japanese woodlander is an exception. A shade lover, it thrives in humus rich moist soil and produces huge heads of rich blue flowers June-Aug. It clumps gently and has an entirely non-invasive habit. Unfortunately slugs find it tasty. Ht. 30cm. Sp. 30cm. **N. *s. Sweet Dreams*.** A lovely sugar pink version of this wonderful woodland plant. **N. *s. Nimbus*** paler blue flowers than the the species.
OMPHALODES. (Boraginaceae). The Greek word 'omphalos' means navel and 'odes' means like. The name refers to the supposedly navel like appearance of the plant's seed

Scabiosa atropurpurea *(3)*

heads. O. *cappadocia Cherry Ingram* has bright blue flowers – much larger than those of the species. May-June. Ht. 25cm. Sp. 40cm. One of very best blue flowered plants. **O. *c. Starry Eyes.*** Large gentian blue flowers with a white star in the centre of each flower. May-June. Bushy plant. Non invasive. A very choice plant. Ht. 18cm. Sp. 35cm. Looks good with Ranunculus gramineus. **O. *verna*** was intrduced from Southern Europe before 1722. It thrives in moistish woodland, and makes dense good ground cover for not too dry shady places. It has deep green, heart shaped foliage and intense, sky blue flowers in April-May. Ht. 10cm. Sp. 45cm. Marie Antoinette was, supposedly, very fond of this unassuming plant. Alice Coats writes "indeed its air of sophisticated innocence seems well suited to the mock shepherdesses and milkmaids of the Trianon." Such royal favour did not protect it from the disdain of Reginald Farrer, who called it the scullion of its race, preferring the more sophisticated charms of O. cappadocia. ***O. v. Alba*** has chalk white flowers. O. verna looks good with Primulas.
PODOPHYLLUM. (Berberidaceae). **P. *hexandrum.*** Amazing marbled leaves appear in May followed quickly by exquisite white, bowl-shaped flowers. Red fruits in summer. Ht. 80cm. Sp. 25cm. Moist rich soil. Shade.
POLYGONATUM. (Asparagaceae). The name comes from the Greek for many, 'poly', and 'gonu' meaning joint. Gerard tells us "the roote is white and thicke and full of knobs or joyntes which in some places resemble the marke of a seale, whereof I thinke it tooke the name sigillum Salamonis". A few pages later he changes his mind and tells us it was called Solomon's Seal, "because of the singular virtue it has in sealing up broken bones and such like" - a very mediaeval appeal to the doctrine of signatures. Solomon's seal is a plant of open woodland in Europe and Asia. Its beautiful, arching shape is best appreciated at the front of the border or in low woodland plantings of Primulas and the like. May-June flowering. In autumn the dark blue, waxy berries are as decorative as its flowers. A fine old Dorset name for this elegant plant is Sow's Tits. Thrives in any not too dry soil. **P. *biflorum* syn P. commutatum.** Greenish white flowers. May-July. Ht. 130cm. Sp. 25cm. **P. *falcatum Variegatum.*** 25cm. tall with a spread of 20cm; has pinkish stems and rounded leaves neatly edged with cream. **P. *hookeri.*** 10cm. tall with a 10cm. spread. **P. *multiflorum.*** Ht. 90cm. Sp. 70cm. **P. *odoratum.*** Quite why it took me 12 years to get this wonderful native plant into our list I do not know. Green and white scented flowers May. Rich dark evergreen leaves. Ht. 50cm. Sp. 20cm.
PRATIA. (Campanulaceae). **P. *pedunculata.*** Starry pale sky blue flowers on dense carpets of tiny green leaves. Ht. 5cm. Sp. 30cm. Looks lovely running out from the edge of a flowerbed into a lawn. June-Sept. Not too dry soil. ***P. p. Country Park.*** Similar habit but deep blue flowers.
PRIMULAS. (Primulaceae). All the following Primulas are easy plants for bog or the edge of a pond or stream, where they will gently self seed and colonise. They are also good plants for sticky heavy clay. **P. *bulleyana.*** Crimson flowers that fade to orange. June-July. Ht. 60cm. Sp. 20cm. **P. *japonicas, P. j. Miller's Crimson*** has dark red flowers and ***P. j. Alba*** has white flowers. They both grow to about 60cm. in height and 20cm. in spread and flower April-May. **P. *florindae***, the giant Himalayan cowslip is a plant, which demands to be the subject of botanical illustration. Its long powdered stems and drooping, soft yellow, deliciously scented flowers are elegance themselves. It flowers throughout the summer months, when all the other Primulas are finished. July-Sept. Ht. 90cm. Sp. 25cm.

Sempervivum arachnoideum *(3)*

P. *polyanthus Gold Lace.* Darkest crimson, gold edged flowers. March-May. Elegant. Ht. 25cm. Sp. 20cm. Moist leafy soil.
P. *veris.* "Cussloppe" (meaning Cow dung) was the Old English name for P. veris; the name derived from the belief that wherever a cow lifted its tail a cowslip sprang up. Presumably it was our genteel Victorian ancestors who prettified the name. Old herbals suggest an infusion of the flowers as a cure for palsy. Cowslip wine is truly delicious. P. veris thrives in rich heavy soil and thin poor soil will need enriching – so don't spare the cowpat! They flower in March April. Deliciously scented Ht. 20cm.
P. *vulgaris.* The Wild Primrose remains the most beautiful of all spring Primulas and perhaps the only spring Primula to look appropriate in the garden. The double Primulas are too ostentatious to mix well in the spring border and seem more at home swanking in pots and window boxes. The wild primrose with its exquisite delicacy looks perfectly at ease with Pulmonarias, Hellebores, Corydalis, Dicentras, Trilliums and spring bulbs. Its pale yellow flowers massed on a bank seem like a reflection of spring's pale sunshine. Ht. 25cm. Sp. 25cm. **Double Primroses.** Double Primulas are rather too flamboyant for the woodland garden. I like to grow them in pots to bring them into the house in flower. They look superb massed in a large bowl as a central piece on a white tablecloth for some special feast. **P. *v. Dawn Ansell*** has double white flowers in a green hose. **P. *v. Miss Indigo*** has deep blue double flowers, laced with silver. **P. *v. Val Horncastle*** has sulphur yellow double flowers. All have a height of 15cm. and a spread of 20cm. They flower Feb-March. Double Primulas are not difficult to grow. They are grandchildren of the ordinary English wild primrose, Primula vulgaris and have the same simple needs. All forms of P. vulgaris require moisture, light shade and reasonably fertile soil. They flourish in clay or leafy woodland soils. They do not thrive in thin poor soils, which should be enriched with plenty of organic matter. Primroses are completely hardy, but enjoy protection from cold wind as this has a desiccating effect from which they are slow to recover. **P. *sieboldii*** is a native of Japan. It is a plant for moist woodland. In Japanese gardens it is often planted under flowering cherries. Its emergent foliage in February is most decorative, pale green and very crimped. In April-May it bears delicate, fragile looking flowers, which are all pastel in colour, pink and blue and white. Ht. 15cm. Sp. 25cm. In humus rich, moist soil it spreads gently. Very reliably perennial.
PULMONARIA. (Boraginaceae). The Latin word signifies suffering from lung disease. The association with lung disease was suggested by the spotted foliage of Pulmonaria officinalis; the mediaeval theory of signatures interpreted this as a sign from God that Pulmonarias had therapeutic powers; hence also the English name lungwort. For the followers of Bacchus rather than Hippocrates, the leaves of Pulmonaria make a very acceptable substitute for borage in Pimms. Pulmonaria officinalis is a European native and has been cultivated in England since mediaeval times. P. longifolia grows wild in parts of Dorset. Pulmonaria angustifolia was introduced from Europe by 1823. Pulmonarias belong with Hellebores, Primulas, Erythroniums and Trilliums in the woodland garden. They are not difficult plants and thrive in any soil, which does not dry out in summer or become waterlogged in winter. I find the Pulmonarias belonging to the rubra group (apart from P. David Ward), are more drought tolerant than others. If cut down after their first flush of flowers, Pulmonarias will oblige with a second flowering. They all make handsome mounds of foliage and flower in March-May. The flowering stems are about 25cm. in height. Pulmonarias make clumps about 30cm. in spread, apart from the P. rubra group, which make bigger plants, perhaps 60cm. in

Sisyrinchium angustifolium (3)

spread. Because the origins of some of the hybrids are obscure, I have placed all the Pulmonarias, regardless of group, in a single alphabetical order with the name of the group, if known, in brackets. **P. *Blue Ensign*** *(angustifolia)* has plain, dark green leaves and large, deep blue flowers. **P. *Cotton Cool*** has violet blue flowers which fade to pink. Its completely silver leaves remain looking pristine throughout the summer. Lovely with Asarum splendens, **P. *David Ward*** *(rubra)* has sage green and silver variegated leaves and coral pink flowers. Needs moist shade, leaves burn if soil allowed to get dry. **P. *Diana Clare*** *(longifolia)* has large violet blue flowers and long pointed silver apple green leaves. Vigorous and foliage looks very good in summer, autumn and winter. **P. *Munstead Blue*.** True blue flowers. Deep green leaves. **P. *Opal*** has deep green foliage with large, silver spots and pale blue flowers; its flowers have a unique elegance and freshness. **P. *Redstart*** *(rubra)* has coral red flowers and plain, apple green foliage; a very vigorous plant; its coral red flowers look particularly well with wild primroses **P. *Sissinghurst*** *(officinalis)* has stunning, pure white flowers very elegantly set off by silver spotted foliage.

RHEUM. (Polygonaceae). Rheum comes from the Greek word 'rheon' meaning rhubarb. Our common rhubarb has some very decorative cousins. **R. *palmatum Ace of Hearts*** has dramatic, jagged, reddish purple foliage. A compact form, it only grows to 100cm. high. Sp. 50cm. It has spikes of pale pink flowers in May-June. Its desiccated flowering spikes still look ornamental well into autumn. **R. *p. var. tanguticum*.** Architectural giant. Spires of pinkish red flowers in June. Purplish red leaves, which fade to green after flowering. Ht. 180cm. Sp. 100cm. Sun or part shade. Rheums need moist fertile soil, but dislike being waterlogged. They look handsome next to water and in smaller gardens take the place of Gunnera. They are happy in sun or part shade. Beware of slugs.

RODGERSIAS. (Saxifragaceae). Rodgersias get their name from the 19th Century Admiral John Rodgers, who commanded the expedition on which they were discovered. Natives to China and Burma, they are definitely not plants for the arid heath lands of Suffolk; they thrive in rich moist soil and luxuriate in the bog garden or in the damp, silty margins of large ponds. Rodgersias are not grown for their flowers, which are somewhat undistinguished. Rodgersias' true magnificence is expressed in their foliage. **R. *aesculifolia*,** as its Latin name suggests, has leaves like that of the Horse Chestnut tree; its large mounds of crinkled bronze leaves make a superb foil for the spear shapes of the larger, water loving irises, such as Iris ensata. White or pink Astilbe like plumes of flowers in July-Aug. Ht 100cm. Sp. 100cm. **R. *tabularis* syn. Astilboides tabularis,** has huge, uncut, round mat green leaves. White Astilbe like flowers in July-Aug. Ht. 100cm. Sp. 200cm. Beth Chatto in her book "The Damp Garden" describes it as one of the most beautiful of all foliage plants. Looks good with the tall spires of Lythrums. Christopher Lloyd suggest planting Rodgersias with hostas, to which they would add something bolder in scale, without subverting them with gaudiness.

ROSCOEA. (Zingiberaceae). **R. *beesiana Alba*.** Thick, clumping, upright stems sheathed in long wavy edged green leaves. Creamy yellow orchid like flowers. May-Aug. Ht. 35cm. Rich moist soil. Part shade. **R. *procera*.** Purple hooded exotic flowers. June-Aug. Glossy green leaves. B. Chatto suggests planting with Tiarellas and Ferns. Ht. 50cm. Sp. 30cm. Does well in heavyish soil.

SANGUINARIA. (Papaveraceae). **S. *canadensis*** takes its name from the Latin word 'sanguis', a reference to the red latex present in its root. Its English name is Blood Root.

Viola Freckles (2)

A member of the poppy family, it is native to Canada and needs moist, humus enriched, woodland conditions. It bears white, cup shaped flowers with golden stamens, which are swiftly followed by deeply cut, pale grey leaves. Summer dormant. Very hardy. Ht. 15cm. Sp. 25cm.

SAXIFRAGA. (Saxifragaceae). **S. *fortunei.*** These plants have been bred in Japan. Valuable for both the freshness of their flowers and their foliage in the late autumn. They like rich moist soil and thrive in shade **S. *f. Blackberry and Apple Pie*** has biscuit coloured flowers and apple green foliage, which is beetroot red on the reverse. Ht. 35cm. Sp. 35cm. **S. *f. Black Ruby*.** Shiny, scalloped black bronze foliage. Red flowers Oct-Dec. Ht. 20cm. Sp. 35cm.

SMILACINA. (Asparagaceae) **syn. Maianthemum. S. *racemosa.*** A Solomon Seal relative and recognisable as such by its arching habit. Its flowers, however, take the form of fluffy cream spikes more in the manner of an Astilbe but far more refined and deliciously scented. May-June. Reginald Farrer describes the subsequent red berries as "vitrified drops of bright blood." Likes a cool moist acid or neutral soil. Ht. 60cm. Sp. 25cm.

SOLDANELLA (Primulaceae). **S. *alpina*** bears delightful fringed, red-flecked, lavender blue flowers in April-May. Ht. 15cm. Sp. 20cm. Likes cool moist, well-drained soil. Soldanellas need shade but dislike being planted under trees or shrubs. To maintain plant vigour they should be lifted and divided after flowering. Always replant into fresh soil. Soldanella flowerbuds are formed in autumn and need protection from slugs over winter.

SYMPHYTUM. (Boraginaceae). **S. *azureum*** . Sky blue flowers April-May. Ht. 50cm. Sp. 40cm. **S. *grandiflorum Goldsmith.*** Dark green leaves variegated with gold. Blue and white flowers in April-May. Ground cover for moist or dry shade. Ht. 15cm. Sp. 40cm. **S. *rubrum.*** A wonderful non-invasive comfrey with deep crimson flowers, which "hang in little crosiers above hairy green leaves." (Graham Stuart Thomas). June-Aug. Ht. 40cm. Sp. 30cm.

TIARELLA. (Saxifragaceae). Tiarellas are members of the saxifrage family. They are evergreen and provide good ground cover in difficult shady areas. They will grow in any soil except bog or pure sand. They all grow 25cm. high with a 25cm. spread. **T. *Mint Chocolate*** has maple shaped, mint green leaves with a large chocolate zonal patch. The edge of the leaf turns bright pink in winter. Toffee coloured flowers in April-Dec. **T. *Ninja*** has fingered palmate leaves with a central chocolate maroon star. Spikes of coral beige flowers in April-June. **T. *Skid's Variegated.*** Pink/green leaves (pinkest in winter) completely covered in cream speckles. Beige white starry flowers. March-May. Ht. 25cm. Sp. 20cm. **T. *wherryi Bronze Beauty*** has maple shaped, pinky leaves. Pale pink, starry flowers. April-Dec.

TOLMIEA. (Saxifragaceae). **T. *menziesii Taff's Gold*** is often grown as a houseplant, but is perfectly hardy in the garden. A native of North West America, it needs to be sited in shady places or the foliage will burn. The foliage is fresh green, delightfully freckled with primrose. In habit like a Tiarella. Brown flowers. May-June. Ht. 30cm. Sp. 30cm.

TRICYRTIS. (Liliaceae). The Toad Lily. Native to Japan, Tricyrtis is one of the most elegant of autumn flowering woodlanders. All the following varieties flower Sept-Oct and are stoloniferous making large clumps in good soil. Ht. 75cm. Sp. 40cm. **T. *formosana*** has glossy dark green, lanceolate leaves, which are speckled with black and pale mauve flowers, heavily spotted with deep purple. **T. *hirta Tojen*** has large lilac purple flowers with white centres. Sept-Nov. Ht. 58cm. Sp. 25cm.

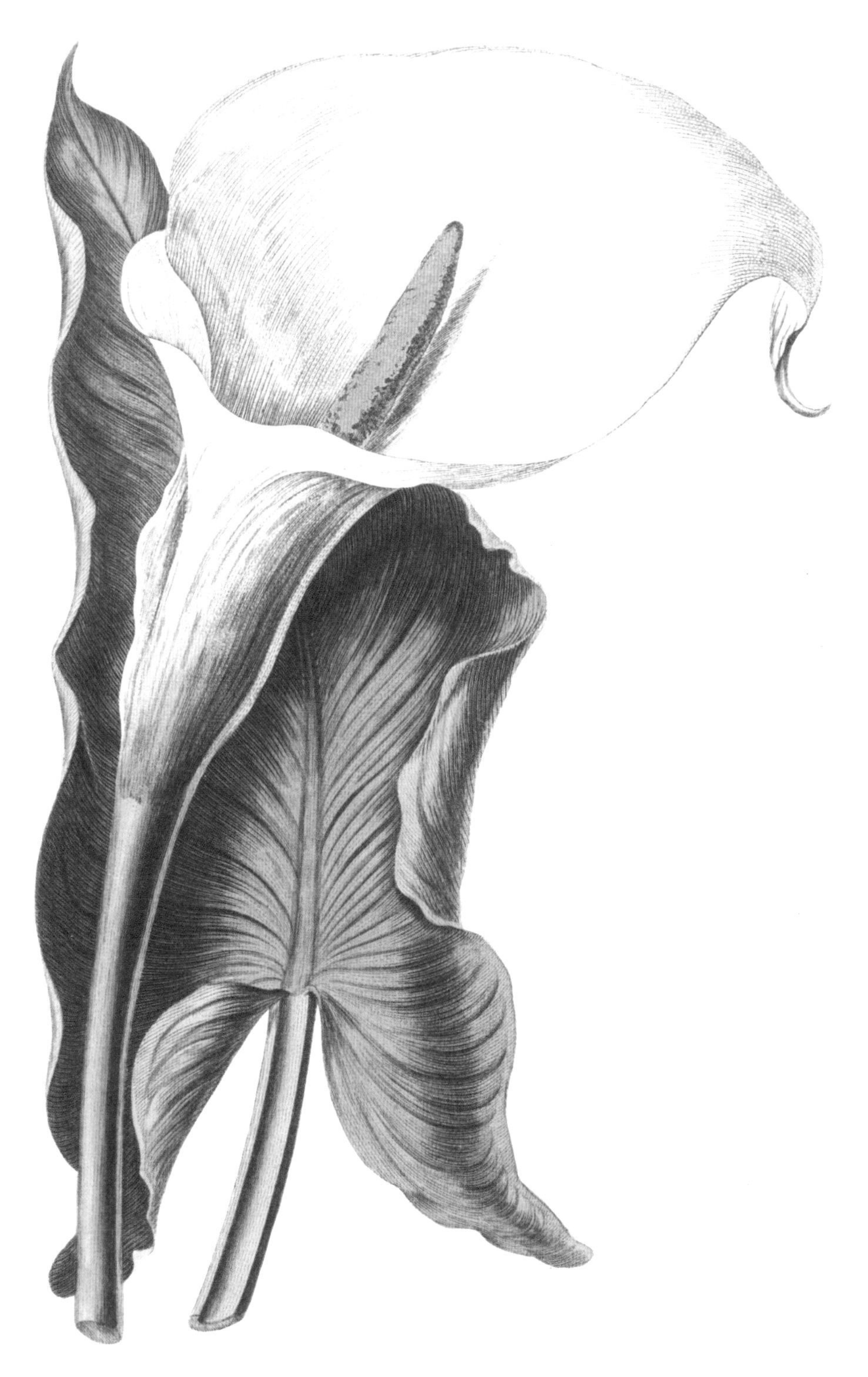

Zantedeschia aethiopica (3)

T. *latifolia White Towers* has unspottted, felted green leaves and unblemished, creamy white flowers. Tricyrtis have branched stems, which carry many small upturned flowers in autumn. Cut one flowering stem and put it in a narrow glass vase and you will find yourself itching to draw it. On the wall in the evening an early autumn fire will etch the plant's profile in riveting shadow. Toad lilies need moisture and light shade and are generally free of pests and diseases. But of slugs beware!

UVULARIA. (Cochicaeae). Uvularias are members of the lily family and related to Solomon's Seal. Native to the Eastern states of North America, they enjoy moist cool woodland conditions. Spring flowering, they become dormant in summer and need careful marking. Stoloniferous, in time they will make large clumps. **U. *grandiflora*** has arching stems and bears pendant, straw yellow flowers in April. Fresh green foliage. Ht. 70cm. Sp. 25cm.

VINCA. (Apocynaceae). Vincas used to be disregarded at Woottens. I think I judged them by their rather miserable performance in our poor dry sand, but, having recently swapped our sand for a handsome loam, with much gnashing and grinding of J.C.Bs, and inspired by a wonderful plate in John Nash's "Garden Flowers", I have decided to give them another try. Vincas derive their name from the Latin verb 'vincere' which has two distinct meanings: 'to conquer' and 'to bind'. Vinca major is definitely a conqueror – and a thuggish and unruly one at that. Vinca minor one need not fear; she gently binds one in her arms. In mediaeval Italy vinca was used to make garlands for condemned criminals from which practice Vinca gained the name 'fiore di morte'. Vinca minor is a native of Britain and has been cultivated in English gardens since the earliest times. Chaucer gives it frequent mention. All varieties of V. minor flower from March through to May. They grow about 20cm. tall with a spread of about 35cm. **V. *minor Atropurpurea*** has deep purplish violet flowers. **V. *m. Bowles Variety*** has large deep blue flowers. **V. *m. Gertrude Jekyll*** has white flowers and is very free flowering. **V. *m. Illumination*.** Green leaves with bright yellow splash. Blue flowers. **V. *m. Marie*** - blue and floriferous. **V. *major Variegata*** is a bold evergreen plant for ground cover in shade. Green leaves splashed with cream. Blue flowers. March-May. Its stems root where they touch the soil. Also good in tubs for winter colour. Ht. 20cm. Sp. 60cm. **V. *major Wojo*.** Bright yellow foliage. Blue Flowers. March-May. Ht. 20cm. Sp. 60cm. **V. *difformis Jenny Pym*.** All the cultivars of V. difformis flower from late autumn into spring. Jenny Pym has pinkish purple flowers with a white stripe. Ht. 30cm. Sp. 60cm.

VIOLA. (Violaceae). The viola cornutas were dealt with in the section on sun loving plants, but other violas thrive in shade. **V. *Freckles*** has white flowers all speckled with blue. **V. *septentrionalis*,** a plant of the most exquisite beauty, has large, blue veined, ivory white flowers. Both plants make neat little clumps of glossy, deep green foliage and flower March-April. Ht. 15cm. Sp. 30cm. The leaves of **V. *labradorica purpurea*** are almost black and act as a superb foil for its blue flowers; it will grow in the driest of dry shade. Ht. 15cm. Sp. 20cm. Beth Chatto suggest planting with Millium effusum aureum and Snowdrops. All these violas flower in spring, but have decorative foliage, which remains attractive throughout the summer.

ZIGADENUS (Liliaceae). **Z. *elegans*.** Pale green star shaped flowers. July-Aug. Dark green, strap shaped leaves. Ht. 35cm. Sp 15cm. Bulbous. Poisonous. Part shade. Moisture retentive soil.

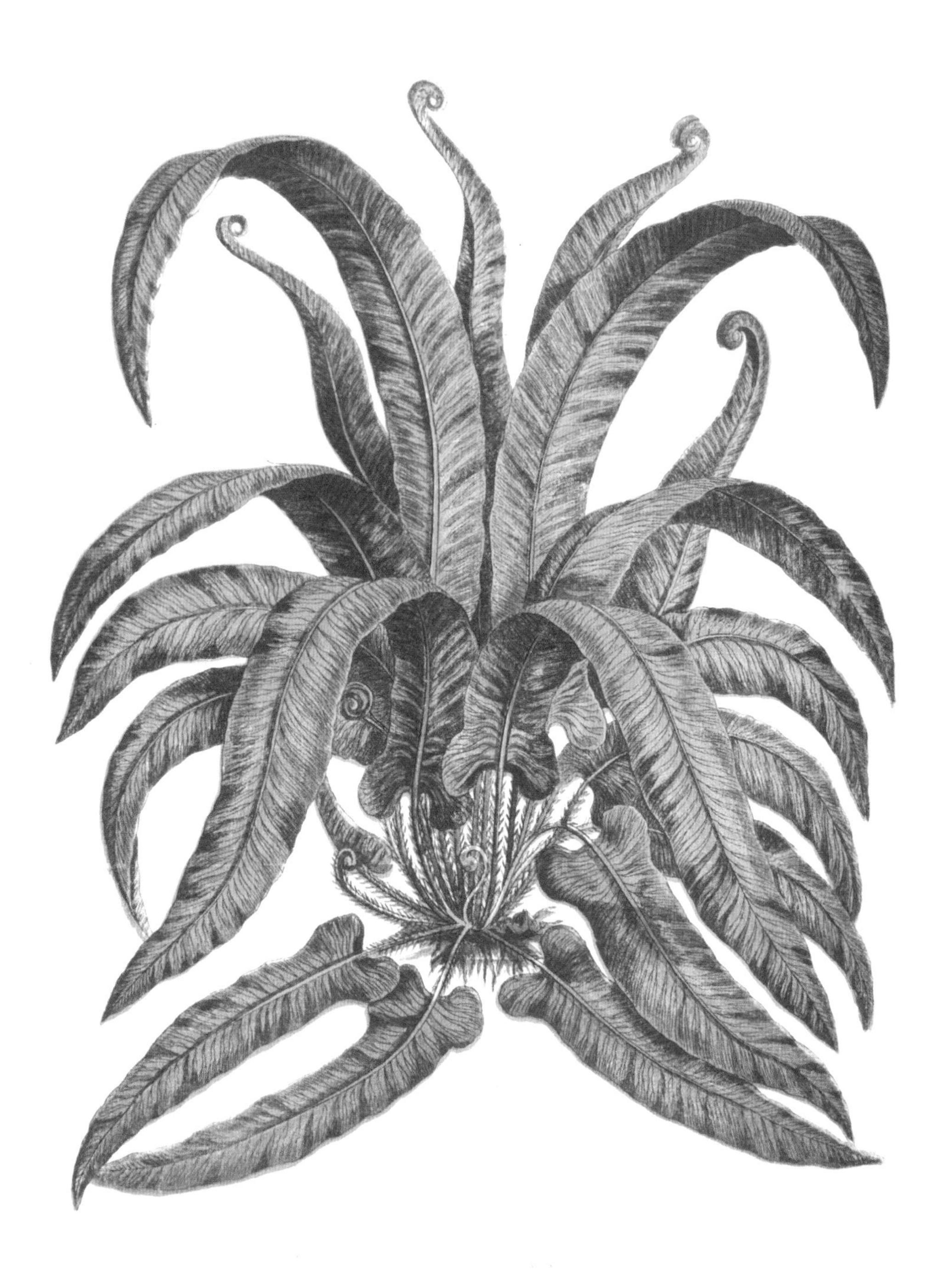

Athyrium scolopendrium (7)

Ferns

The following list is only a tentative beginning. Having put together a fairly respectable collection of Hostas, it seemed essential to be able to offer some ferns. They not only, in the main, delight in the same conditions - moist cool shade, but are superb visual complements; Ferns with their delicate divided leaves dance against the altogether more substantial foliage of Hostas. All these ferns are lime tolerant.

ASPLENIUM. (Aspleniaceae). **A. *scolopendrium.*** The hart's tongue fern differs from most ferns in not having divided foliage. It has rich green glossy strap like leaves. It grows best in moist, but not wet, alkaline shade. Looks stunning with Carex elata Aurea. Ht. 60cm. Sp. 60cm.

ATHYRIUM. (Woodsiaceae). **A. *nipponicum Pictum.*** The Japanese painted fern has metallic, greyish green leaves with purple midribs. Moist shade. Acid or alkaline soil. Ht. 35cm. Sp. 35cm. **A. *nipponicum Ursula's Red.*** Reddish pink and silver fronds develop a central black stripe as they mature. Ht. 30cm. Sp. 30cm.

BLECHNUM. (Blechnaceae). **B. *penna marina.*** Evergreen New Zealand hardy ground cover. Narrow deep green fronds. Ht. 20cm. Sp. 40cm. Moist or dry soil. Christopher Lloyd recommends it for planting in paving in cool shade. **B. *spicant.*** Evergreen fern. Narrow lance shaped fronds. Glossy, dark green leaves. Ht. 35cm. Sp 40cm. Dry or moist, not too alkaline soil. Shade or part shade.

DRYOPTERIS. (Dryopteridaceae). All Dryopteris tolerate a wide range of conditions thriving in both acid and alkaline, moist and dry soil. They all require some shade. **D. *affinis,*** the Golden Male Fern, has dark green fronds with golden stems. Ht. 100cm. Sp. 80cm. **D. *erythrosora.*** Fronds emerge pink in spring, turning bronze as they mature, finally in winter colouring yellow. Ht. 60cm. Sp. 60cm. **D. *filix mas,*** The Common Male Fern, is a robust easy plant with fresh green fronds. Ht. 100cm. Sp. 80cm. **D. *filix mas linearis.*** Dark green fronds with very narrow leaves. Very different. Elegant and tough. Ht. 60cm. Sp. 60cm.

MATTEUCCIA. (Woodsiaceae). **M. *struthiopteris.*** Fronds of pale yellow-green foliage form an upright chalice, hence its English name, the Shuttlecock Fern. Moist shade. Ht. 90cm. Sp. 70cm.

OSMUNDA. (Osmundaceae). **O. *regalis.*** "The Royal Fern." An upstanding giant of a fern. An established specimen will reach 120cm. in height. Likes acid or neutral, moist rich soil. Also thrives in bog, but does not like to have its crown submerged. Cool spot. Sp. 45cm.

POLYSTICHUM. (Dryopteridaceae). **P. *polyblepharum.*** Attractive golden-bristled croziers in early spring, from which emerge yellowish-green lance-shaped evergreen fronds; these turn deep green later. Any not bone dry soil. Shade. Ht. 45cm. Sp. 60cm. **P. *setiferum.*** Soft shield fern. Evergreen lacy foliage. Ht. 90cm. Sp. 80cm. Any well drained soil in shade. Likes lime. **P. *setiferum Plumosum Densum.*** A very finely cut form of this elegant fern. Ht. 50cm. Sp. 40cm. Any not bone dry soil. Shade.

Blechnum spicant (7)

Hostas

HOSTA (Asparagaceae). Hostas thrive in any well-worked moisture-retentive soil. Yellow leaf varieties should be grown in full sun; in shade they will fail to develop their full colour. Green leafed varieties are grown primarily for their flowers and should also be planted in full sun, as in shade their flowers may fail to mature before the frosts. Blue leafed varieties scorch easily and should be planted in deep shade. Varieties with golden variegation are best planted where they will receive some gentle sunlight in the early morning or late afternoon - give them too much sun and they will scorch, too little and the variegation will fail to develop. Varieties with white variegation can be planted in dappled or deep shade. Hostas are generally free from pests and diseases with the exception of slugs and snails. Slug pellets should be first put out just before the foliage buds emerge through the soil, as slugs are educated to drill neat holes through each bud: the leaves unfold and each and every leaf bears witness to your negligence for the rest of the season. If Hostas are grown in pots they should be treated every autumn as vine weevil can be a problem. If you are too idle to protect your Hostas you may be interested to know that the following hostas show some resistance to slugs. **Slug resistant blue varieties:** Blue Angel, Blue Moon, Krossa Regal. **Slug resistant yellow varieties:** Sum and Substance. **Slug resistant green variety:** Invincible. **Slug resistant variegated varieties:** Northern Exposure, Queen Josephine. Hostas make excellent troublefree plants for large pots or tubs. They are perhaps best planted as a single variety to each container. The bold simplicity of their foliage will bring tranquility and restore calm in a world of blather.

Yellow leafed varieties. All yellow leafed Hostas should be planted in full sun. In shade they will not develop their true leaf colour. **H. *Golden Sunburst.*** Chartreuse leaves which become bright yellow. Ht. 45 cm. Sp. 90cm. Palest lavender flowers. June-July. Vigorous. **H. *Sum and Substance.*** A big beast among hostas. Huge 75cm. high leaves. Sp. 150cm. Bright yellow in full sun. Chartreuse green in shade. Pale lavender flowers mid season. Sometimes repeat flowers in autumn. Has the reputation of being slug resistant.

Blue leafed varieties. The blue leafed Hostas should all be planted in deepish shade. In sunlight their foliage will scorch. **H. *Big Daddy*** (sieboldiana group) has almost round, blue, puckered leaves. Slow growing. White flowers in July-August. 60cm. **H. *Blue Angel.*** White flowers. Blue leaf. Ht. 105cm. Slug resistant. **H. *Blue Cadet*** (tokudama group) has small grey blue leaves and purple flowers excellent for edging. Palest lavender flowers. June-July. Increases readily. Ht. 35cm. Sp. 30cm. 14ins. **H. *Blue Moon.*** The smallest of the tardianas. Intense blue leaves. Palest lavender flowers. June-July. Ht. 20cm. Sp. 20cm. Slug resistant. **H. *Halcyon*** (tardiana group) has small striking blue grey leaves. Greyish lavender flowers in June-July. Ht. 35cm. Sp. 35cm. Slug resistant. **H. *Krossa Regal*** (sieboldiana group) has large grey blue leaves on long stems. Lilac flowers in August. Ht. 90cm. Sp. 90cm. **H. *Love Pat*** (tokudama group). Intensely blue, deeply cupped, puckered leaves. White flowers with the palest tinge of lavender in June-July. Ht. 60cm. Sp. 60cm. **H. *sieboldiana Elegans*** has huge cup shaped, puckered, blue-grey leaves. Lilac flowers in June-July. Ht. 75cm. Sp. 75cm.

Dryopteris felix mas (7)

Green leafed varieties. H. *Devon Green*. Tardiana group. Small green leaves. Ht. 50cm. Sp. 50cm. Palest lavender flowers. June-July. **H. *Invincible*** Olive green leaves with very shiny undersides. Scented lavender (occasionally double) flowers. Late flowering. Ht. 60cm. Slug resistant.

Variegated leaf varieties. H. *abiqua Moonbeam*. Sport of August Moon. Leaves emerge plain mid green, and gradually acquire a chartreuse edge. Best in some sun. Pale lavender flowers in June-July. Ht. 60cm. Sp. 60cm. **H. *flavocircinalis*.** Tokudama group. Oval leaves with irregular yellow margin. Palest lavender flowers. June-July. Ht. 45cm. Sp. 45cm. **H. *fortunei Fire and Ice*** has pure white leaves with deep green margins. Lavender flowers in July. A very dainty, distinctive plant. Ht. 30cm. Sp. 30cm. Light to full shade. **H. *fortunei Gold Standard*** has large leaves, which emerge green, later turning gold with a deep green margin. Lavender flowers in June-July. Ht. 50cm. Sp. 50cm. **H. *Frances Williams*** has large, grey blue leaves with a creamy yellow edge. Flowers palest lavender. June-July. Magnificent, when well grown, but needs full shade as sun will scorch its chartreuse edge. Pale lavender flowers in July. Ht. 75cm. Sp. 75cm. **H. *Inniswood*** has very large, round golden foliage with a wide, green edge. Lavender flowers in June-July. Light or part shade. Ht. 60cm. Sp. 60cm. **H. *Karin*.** Dark green leaves with a creamy white edge. Ht. 40 cm. Sp. 40cm. Part or full shade. Any not too dry soil. **H. *Loyalist*.** Deep green margins with a pure white centre. Lavender flowers. July-Aug. Ht. 55cm. Sp. 100cm. **H. *Minuteman*** has thick, cupped, blue green leaves with a wide white margin. Lavender flowers in June-July. Ht. 50cm. Sp. 40cm. **H. *Night Before Christmas*** has pointed, twisted, white leaves with deep green margins. Lavender flowers in June-July. Ht. 35cm. Sp. 35cm. **H. *Northern Exposure*.** Large puckered blue leaves with a wide creamy yellow margin. White flowers in June-July. Ht. 50cm. Sp. 50cm. Slug resistant. **H. *Patriot*.** Puckered olive green leaves with very wide white margin. Diana Grenfell thinks it "probably the best white edged Hosta ever raised." Lavender flowers. Ht. 55cm. Sp. 100cm. Light to full shade. **H. *Paul's Glory*** has yellow leaves with a bluish green margin. In late season the yellow centres become creamy white. Lavender flowers in June-July. Ht. 40cm. Sp. 35cm. Light shade to full sun. **H. *Queen Josephine*.** Deep green, shiny foliage with a golden margin. Lavender flowers. June-August. Ht. 30cm. Sp. 30cm. Light to full shade. **H. *sieboldiana Maple Leaf*.** Blue green leaves with yellow edge. Palest lavender flowers. June-July. Full shade. Ht. 40 cm. Sp. 40cm. **H. *So Sweet*.** As its name suggests is a very sweetly perfumed hosta. Mauve flowers mid to late season Glossy green leaves with a creamy margin. Ht. 35cm. Sp. 55cm. **H. *Summer Music*** has heart shaped leaves with white centres and deep green margins. Light lavender flowers in June-July. Full shade. Ht. 50cm. Sp. 30cm.

Osmunda regalis

Credits

Design, Sally Seeley, Drab Ltd. 2005 and 2013.

Dust Jacket Photo, Viv Kemp 2004.

(1) Curtis's Botanical Magazine. London 1789-1870.

(2) Christine Stephenson botanical illustrations commissioned by Woottens. 1994-2004.

(3) New Botanic Garden. London 1812.

(4) Paxton's Magazine Of Botany. London 1838-1847.

(5) The Florist's Directory by James Maddock Printed by John Green. London 1822.

(6) English Garden Flowers John Nash Hodder. London 1948.

(7) Flowering Plants Of Great Britain by Ann Pratt. F Warne and Co London 1874.

Plants for Problem Areas

The lists below are just a guide. For a comprehensive search, which enables you to find plants to suit both your conditions and tastes, we do recommend you consult our website www.woottensplants.co.uk

Plants for Dry Shade

Gardening books tend to be overoptimistic about recommending plants for dry shade. All the plants listed below will grow in almost pure sand under deciduous trees. This list owes much to one produced by Beth Chatto in her handbook, "Unusual Plants". *Aegopodium. Aquilegia vulgaris. Allium oreophilum. Arum italicum Pictum. Aster divaricatus. Aster novi belgii Twilight. Astrantia major. Anchusa sempervirens. Bergenia. Blechnum penna-marina. Carex buchananii. Cyclamen. Cyrtomium fortunei. Digitalis purpurea. Dryopteris. Epimedium. Eupatorium rugosum. Geranium macrorrhizum. Geranium nodosum. Geranium phaeum. Geranium sylvaticum. Helleborus foetidus. Helianthus Lemon Queen. Hostas. Iris foetidissima. Lamium. Lilium philipinense. Luzula. Milium. Polygonatum. Polystichum setiferum. Pulmonaria rubra. Symphytum. Tellima grandiflora. Tiarella. Tolmeia. Vinca. Viola labradorica.*

Plants for Dry Soil and Sun

Acanthus. Achillea. Agastache. Alchemilla. Allium. Anisodontea. Anthemis. Aquilegia. Arthropodium candidum purpureum. Anthriscus. Argyranthemum. Armeria. Artemisia (all except Artemisia lactiflora). Asphodeline. Astelia. Aster divaricatus, Aster novi belgii Twilight. Bergenia. Berkheya. Blechnum penna-marina. Briza media. Buddleja davidii. Calamintha nepetoides. Calamogrostis. Campanula alliariifolia. Campanula cochlearifolia. Campanula garganica. Campanula lactiflora. Campanula persicifolia. Campanula poscharskyana. Carex phyllocephalla. Caryopteris. Catanache. Centaurea. Centranthus. Ceratostigma. Cerinthe. Cichorium. Cistus. Convolvulus. Crambe. Crocosmia. Cynara. Dianthus. Diascia. Echinacea. Echinops.

Eremurus. Erigeron glaucum. Erigeron karvinskianus. Eryngium. Erysimum. Euphorbia. Festuca. Filipendula. Foeniculum. Geranium (all). Gaura. Gladiolus. Gypsophila repens. Halimiocistus. Helianthemum. Helianthus Lemon Queen. Helleborus argutifolius. Helleborus foetidus. Hemerocallis. Hermodactylus. Hostas (Yellow and green leafed). Hyssopus. Imperata cylindrica Rubra. Inula. Iris - Bearded. Iris sibirica. Iris unguicularis. Lamium. Lathyrus latifolius. Lavandula. Lavatera. Libertia. Limonium. Linaria. Lippia citriodora. Lithodora. Lobelia Elmfeur. Lupinus. Luzula. Lychnis coronaria. Malva moschata. Matthiola fruticulosa. Miscanthus. Molinia. Myrtus communis. Nemesia. Nepeta (all except N. govaniana and N. subsessilis). Oenothera. Onopordum. Origanum. Osteospermum. Paeonies. Panicum. Papaver (all). Pennisetum orientale. Perovskia. Phlomis. Phormium. Physostegia. Polemonium. Ranunculus Brazen Hussy. Rhodanthe. Rosmarinus. Rudbeckia. Salvia (not S. superba or S. uliginosa). Sanguisorba. Santolina. Saponaria ocymoides. Saxifraga x urbium Aureopunctata. Scutellaria incana. Sedum. Sempervivum. Sesleria. Silene uniflora. Sisyrinchium. Spartina. Sphaeralcea. Stachys byzantina. Stipa (all except S. arundinacea). Succisa pratensis. Thymus. Tradescantia. Tulbaghia. Verbascum. Verbena. Veronica peduncularis. Yucca.

Plants for Irredeemable Clay

Achillea ptarmica The Pearl. Aconitum. Ajuga. Alchemilla. Allium christophii. Althaea officinalis. Amsonia. Anaphalis. Anchusa sempervirens. Anemone x (Japanese Anemones). Angelica. Aruncus. Asters. Astilbes. Brunneras. Camassia. Calamagrostis. Caltha palustris. Campanula lactiflora. Campanula latiloba. Campanula latifolia. Campanula Kent Belle. Carex. Chelone. Chrysanthemums. Cimicifuga actaea. Dicentra formosa. Eupatoriums. Euphorbia robbiae. Filipendula. Galega. Geranium Dusky Crug. Geranium macrorrhizum. Geranium phaeum. Geranium sinense. Geranium sylvaticum. Geum. Hakonechloa. Helleborus niger. Hemerocallis. Heucheras. Hostas. Houttuynias. Iris ensata. Iris laevigata. Iris sibirica. Iris spuria. Ligularia. Luzulas. Lychnis chalcedonica. Lobelias. Lychnis flos cuculi. Lysimachia. Lythrum. Miscanthus. Molinias. Monarda. Myosotidium. Omphalodes verna. Panicum. Pennisetum alopecuroides. Persicaria. Phlox. Phygelius. Physostegia. Polygonatum. Primula. Pulmonaria. Ranunculus. Rheum. Rodgersia. Roscoea. Rudbeckia. Sanguisorba. Saponaria officinalis. Saxifraga fortunei. Schizostylis. Thalictrum. Tiarella. Tradescantia. Tiarella. Trollius. Uvularia. Valeriana. Veronica gentianoides. Vinca major. Zantedeschia.

Rabbit Proof Plants

Aconitum. Aegopodium. Agapanthus. Alchemilla. Anaphalis. Anemone x (Japanese Anemones). Aquilegia. Armeria. Asphodeline. Aster. Astilbe. Bergenia. Brunnera. Campanula lactiflora. Campanula latifolia. Centaurea. Clematis. Corydalis. Crinum. Crocosmia. Cyclamen. Cynara. Digitalis. Doronicum. Epimedium. Eupatorium. Euphorbia. Fuchsia. Galanthus. Geranium. Helenium. Helianthus. Helleborus orientalis. Hemerocallis. Hosta. Iris. Kirengeshoma. Lamium. Leucojum. Lysimachia. Malva. Miscanthus. Nepeta. Omphalodes. Paeonia. Papaver. Phormium. Polygonatum. Pulmonaria. Rheum. Rodgersia. Rosmarinus. Saxifraga x urbium. Sedum. Stachys byzantina. Tradescantia. Trollius. Vinca.

Bibliography

E. A. Bowles, 'My Garden In Spring'. Timber Press, Oregon 1998.

E. A. Bowles, 'My Garden In Summer'. Timber Press, Oregon 1998.

E. A. Bowles, 'My Garden In Autumn And Winter'. Timber Press, Oregon 1998.

Beth Chatto, 'Plant Portraits'. Dent, London 1985.

Alice M. Coats, 'Flowers And Their Histories'. Hulton Press, London 1956.

Rick Darke, 'The Colour Encyclopaedia of Ornamental Grasses'. Timber Press, Oregon 1999.

Reginald Farrer, 'The English Rock Garden', vols I and II. Nelson, London 1919.

John Gerard, 'Herball'. Gerald Howe, London 1927.

Geoffrey Grigson, 'The Englishman's Flora'. Helicon, Oxford, 1996.

Jason Hill, 'The Curious Gardener'. Faber, London 1932.

Jason Hill, 'The Contemplative Gardener'. Faber, London 1939.

Michael King and Piet Oudolf, 'Gardening With Grasses'. Francis Lincoln, London 1996.

Christopher Lloyd, 'Garden Flowers'. Cassell, Jersey 2000.

Roger Phillips and Martyn Rix, 'Perennials', vols I and II. Pan, London 1991.

Vita Sackville-West, 'In Your Garden'. Michael Joseph, London 1958.

Vita Sackville-West, 'In Your Garden Again'. Michael Joseph, London 1953.

Vita Sackville-West, 'More For Your Garden'. Michael Joseph, London 1955.

Vita Sackville-West, 'Even More For Your Garden'. Michael Joseph, London 1958.

F. C. Stern, "A Chalk Garden" Faber, London 1974.

G. S. Thomas, 'Perennial Garden Plants'. Francis Lincoln, London May 2004.

Ann Pratt, 'Flowering Plants Of Great Britain'. F. Warne and Co, London 1874.

Flowering Plant Classification from APG III 'Angiosperm Phylogeny Group' 2009.